ECOLOGICAL NICHES

INTERSPECIFIC INTERACTIONS

A Series Edited by John N. Thompson

ECOLOGICAL NICHES

LINKING CLASSICAL AND CONTEMPORARY APPROACHES

JONATHAN M. CHASE AND
MATHEW A. LEIBOLD

THE UNIVERSITY OF CHICAGO PRESS
CHICAGO AND LONDON

JONATHAN M. CHASE is assistant professor in the Department of Biology at Washington University and is the author of several published papers. MATHEW A. LEIBOLD is associate professor in the Section of Integrative Biology at the University of Texas at Austin; he is the author of many published papers and is a coeditor of a new edition of Charles Elton's *Animal Ecology* (2001).

The University of Chicago Press, Chicago 60637
The University of Chicago Press, Ltd., London

Printed in the United States of America

12 11 10 09 08 07 06 05 04 03   1 2 3 4 5

ISBN: 0-226-10179-7 (cloth)
ISBN: 0-226-10180-0 (paper)

Library of Congress Cataloging-in-Publication Data

Chase, Jonathan M.
Ecological niches : linking classical and contemporary approaches / Jonathan M. Chase and Mathew A. Leibold.
p. cm — (Interspecific interactions)
Includes bibliographical references (p.).
ISBN 0-226-10179-7 (cloth : alk. paper) — ISBN 0-226-10180-0 (pbk. : alk. paper)
1. Niche (Ecology) I. Leibold, Mathew A. II. Title. III. Series.
QH546.3.C53 2003
577.8′2—dc21

2002013308

♾ The paper used in this publication meets the minimum requirements of the American National Standard for Information Sciences—Permanence of Paper for Printed Library Materials, ANSI Z39.48-1992.

CONTENTS

PREFACE

Many contemporary ecologists probably received their earliest exposure to the concept of *ecological niches* from Theodor Seuss Geisel's writings:

> and NUH is the letter I use to spell Nutches,
> Who live in small caves, known as Niches, for hutches.
> These Nutches have troubles, the biggest of which is
> The fact that there are many more Nutches than Niches.
> Each Nutch in a Nich knows that some other Nutch
> Would like to move into his Nich very much
> So each Nutch in a Nich has to watch that small Nich
> Or Nutches who haven't got Niches will snitch.
> —Dr. Seuss, *On Beyond Zebra* (1955)

Why does an organism perform as it does? Why does it live where it lives? Why does it eat what it eats? How do organisms interact with one another? Which organisms can coexist with one another? Why are some species abundant and others rare? Why are some species widespread and others localized? What determines how many kinds of organisms will coexist through space and time? How do ecological interactions influence a species' evolutionary trajectories? What consequences does the presence of a species have on ecosystem-level process and function? These questions, and a thousand others, are related to the oft maligned but centrally important concept of a species' ecological niche.

We have two primary goals in this book. First, the niche concept has been central throughout much of the development of ecology, but as it has fallen out of favor recently in deference to more specific conceptual frameworks, many of the ideas and theories that were developed in previous decades are being ignored. We believe that an ecology firmly based on its previous successes and failures will be more successful than one that ignores its history. Thus, we connect many of the pressing questions, theories, and tools of modern ecology to their classical counterparts. Second, ecology has become a very diffuse science. Evolutionary, population, community, and ecosystems ecology have become nearly separate disciplines with little cross-fertilization. We believe that the niche concept lies at the nexus between these disciplines, and that its reformalization will help to synthesize these diverging fields.

Hubbell (2001) has recently challenged ecologists by claiming that the niche concept will have to be completely rethought from first principles. As an alternative, Hubbell champions a neutralist theory of biodiversity and relative species abundance, which claims that understanding differences among species with regard to their niches is not necessary and that many broad and general ecological patterns can be understood by assuming that all species are, in essence, identical in their ecological niches. However, it is well known that species do indeed differ in many important aspects, and so Hubbell's assumption cannot be strictly true. Nevertheless, the apparent success and predictive ability of Hubbell's ideas place the centrality of the niche concept in ecology at serious risk. What we hope is that this book will do just what Hubbell asked. Specifically, we will place the concept of the niche into a less ambiguous framework and try to disentangle a number of complicated processes that have caused confusion and dissatisfaction. In doing so, we hope to help answer a number of unresolved questions and to reestablish the central position of the niche concept in ecology and evolutionary biology.

Chapter 1 provides an overview of the history of the use of the niche concept in ecology and places it in its current context. Chapter 2 uses the very simplest possible models to help us refine the niche concept and illustrate its use in several important situations. Chapter 3 contrasts this with the conventional approach and discusses the general role of trade-offs in ecology. In chapter 4 we discuss the issues involved in the empirical testing and use of these ideas. Chapter 5 relaxes some of the assumptions of the simplified framework and explores some of the effects of adding physiological, behavioral, and life-history complexities. Chapter 6 further relaxes some assumptions by exploring the effects of fluctuating environments and other nonequilibrial situations. Chapter 7 develops the concept of species sorting by extending the use of the niche concept beyond the local scale to the regional scale. Chapter 8 then takes these basic components of the niche and explores some of the larger-scale problems of community ecology, including patterns of succession, community assembly, and species composition and diversity. Chapter 9 examines how the niche concept can be used in the study of ecosystem attributes. Chapter 10 explores some of the effects of evolutionary biology on the dynamics of ecological niches and diversification and shows some ways to incorporate ecological theory into evolutionary studies, and conversely how to incorporate evolution into ecological investigations. We end in chapter 11 with a synopsis and prospectus for approaches using the niche to continue synthetic studies of ecology.

We have many people to thank for helping us along the way. Several years ago, Michael Usher first suggested to us that a book on ecological niches was needed. His advice and comments early on helped us give the project life. Christie Henry, at the University of Chicago Press, provided much-needed encouragement, helpful suggestions, and gentle pushing at all the right times. The staff at the press also provided expert help at crucial stages, particularly Jennifer Howard, for help with intangibles too numerous to note, and Michael Koplow, whose skills as a copyeditor and eye for detail went above and beyond the call of duty. John Thompson, the series editor, provided many helpful suggestions on our writing and on the process throughout. We thank Tom Miller for a very detailed and extremely helpful forty-page single-spaced review of the first draft and another helpful review of the second draft of the entire manuscript. Jonathan Losos and an anonymous reviewer also provided many helpful suggestions on a previous draft of the entire book. Peter Chesson, Bob Holt, and Joel Brown gave many helpful suggestions on several chapters, and Tim Wootton and several students in our graduate classes read and commented on select chapters. Julia Butzler read through the entire manuscript making sure all of the i's were dotted and t's crossed (and figures and text matched, were correctly labeled, etc.), and also provided many helpful suggestions. Finally, Tiffany Knight spent many long nights reading over the entire manuscript with J. M. C., making sure all of the text was readable, understandable, consistent, and correct; this book has benefited enormously because of her hard work. Of course, any inconsistencies, confusion, or mistakes remain our fault. Support for our research, and our collaboration, was provided by the University of Chicago, the University of California-Davis, the University of Pittsburgh, Washington University, and the National Science Foundation. Finally, we thank our families and our spouses Tiffany Knight (J. M. C.) and Eve Whitaker (M. A. L.), whose love, support, and understanding endure.

CHAPTER ONE

INTRODUCTION: HISTORY, CONTEXT, AND PURPOSE

The niche concept remains one of the most confusing, and yet important, topics in ecology.
—R. B. Root (1967)

I think it is good practice to avoid the term niche whenever possible.
—M. H. Williamson (1972)

The niche concept is a very useful addition to the ecologist's tool kit because it combines the best properties of both baling wire and putty; it holds ill-fitting pieces together that would otherwise fall apart and at the same time fills the gaps between them so that poor workmanship may go undetected.
—D. Reilloc (cited in Hurlbert 1981)

At its most ambitious, the theory of niche helps us understand fundamental questions of ecology.
—T. W. Schoener (1989)

No concept in ecology has been more variously defined or more universally confused than "niche."
—L. A. Real and S. A. Levin (1991)

The concept of niche provides a way of characterizing important ecological attributes of species while recognizing their uniqueness.
—J. H. Brown (1995)

Ecology's love-hate relationship with the niche concept has been long and not especially pretty.
—N. G. Hairston Jr. (1995)

Studies of the niche have played an important role in the development of community ecology, and are likely to do so in the future.
—B. A. Maurer (1999)

I believe that community ecology will have to rethink completely the classical niche-assembly paradigm from first principles.
—S. P. Hubbell (2001)

1.1. Clearing the Decks and Setting the Stage

This book is directed at reinterpreting and reevaluating the concept of the niche within the framework of recent developments in ecology. The field of ecology has developed tremendously in recent years as theoretical issues have been resolved, as experimental approaches have been refined, and as new questions and approaches have been developed. In this rapidly shifting conceptual landscape, old ideas can often seem to disappear or change in ways that make them unrecognizable. One of these is the concept of the *ecological niche*, which for now we loosely define as the requirement of a species for existence in a given environment and its impacts on that environment (a more formal definition is at the end of this chapter).

The niche concept is an important element in almost every aspect of ecological thinking, from the study of behavior, morphology, and physiology of individual organisms to approaches that evaluate how species participate in ecosystem functioning. From its inception in the ecological literature, most notably due to Grinnell (1917), Elton (1927), and Gause (1936), use of the niche concept increased steadily in the ecological literature (fig. 1.1). In fact, the term "niche" appeared in more than one-fourth of all papers published in the journal *Ecology* during its peak usage in the 1960s and 1970s. However, over the past twenty or so years, the niche has declined significantly as an ecological concept. The term seems to be almost avoided by many ecologists, and there is a perception that it is a concept that has outlived its utility. We argue to the contrary that many of the concepts and theoretical tools that provide the deepest insights into ecology are intimately linked to the idea of the niche and that a recognition of this link can help to solidify the conceptual synthesis that ecology so desperately needs.

The niche concept we develop in this book is probably valid for almost any situation of ecological interest. A species is consistently able to exist for long periods of time only when its ecological requirements are met in a local environment. "Sink" habitats, in which a species could not exist unless supported by immigration from a highly productive "source" habitat, are a notable and important exception (Pulliam 1988, 2000), but one that begs understanding of how a species' requirements are met at these sources (see also Loreau and Mouquet 1999; Amarasekare and Nisbet 2001). Further, the niche concept provides a richer and potentially more instructive context when it is used to evaluate species interactions in communities. Our approach is based on the idea that differences among species in their niches (either their

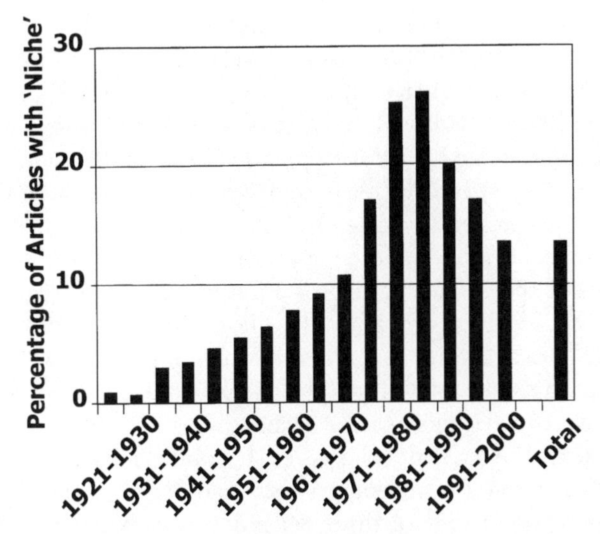

Fig. 1.1. Percentage of all articles in *Ecology* that contained the word "niche." Data from 1920–1995 were obtained from the internet database JSTOR (www.jstor.org) by searching for the keyword "niche" in the full text of articles. Data from 1996–2000 were obtained by performing the same search on the journal's homepage (www.esa.org).

requirements or their impacts or both) are important in determining the outcome of species interactions as might be revealed in the their distributions and/or abundances, as well as in their biodiversity and their functional role in ecosystems. There is a very strong historical legacy (Elton 1927; Hutchinson 1957, 1959, 1978; MacArthur 1958, 1972; Williams 1964; Levin 1970) and contemporary focus (Tilman 1982, 1988, 1999; Chesson 1991, 2000a,b; Leibold 1995; Weiher and Keddy 1999; Pulliam 2000) of the niche concept in ecology. However, there is also a considerable amount of debate as to whether the concept of the niche, with particular regards to local interactions and ecological differences among species, is necessary or useful in understanding broad patterns in the diversity, distribution, and abundance of species (Bell 2001; Hubbell 2001).

Recently, Hubbell (2001) has suggested that ideas related to the niche may not be needed at all, and that the classical paradigm based on the niche perspective needs to be completely rethought. To that end, Hubbell has comprehensively developed an alternative approach, called the "unified neutral theory of biogeography," with the stated assumption that all species are identical with respect to their ecological traits, and to "see how far" he could go with it in explaining natural patterns of species diversity and abundance. Hubbell based his neutral theory on a synthesis of ideas and data from island biogeography (MacArthur and Wilson 1967), previous neutral ecological models (Caswell 1976), his own studies in tropical forests (Hubbell 1979; Hubbell and Foster 1986; Hubbell et al. 1999), and a strong analogy with concepts of genetic drift in population genetics and neutral speciation in evolutionary biology. Although the key assumption of neutrality is not likely to be strictly true, as Hubbell freely admits, the theory does a remarkable job of predicting patterns of biodiversity and relative abundance of species in a variety of different ecosystems (see also Bell 2000, 2001). Furthermore, Hubbell's ideas have reinvigorated theoretical and empirical studies aimed at understanding patterns of biodiversity and relative species abundance at larger spatial scales (Bell 2001; Chave et al. 2002; Condit et al. 2002).

The neutral theory is not, however, beyond criticism. First, species do differ in their traits and often show trade-offs that allow them to coexist for long periods of time. Second, many experimental studies and natural disturbances show that when a system is perturbed away from its equilibrium, it often tends to go back to that equilibrium. Third, several recent theoretical models show that the predictions of the neutral theory can be quite tenuous and highly dependent on unrealistic parameter values (Chesson and Huntly 1997; Zhang and Lin 1997; Chesson 2000b). Fourth, predictions from the neutral model are difficult to distinguish from the predictions of models with niche differences (Chave et al. 2002). Finally, many empirical patterns, even within the very tropical forest systems that Hubbell (Hubbell 1979; Hubbell and Foster 1986; Hubbell et al. 1999) used to develop his ideas, do not conform to many of the neutral model's predictions on patterns of biodiversity, species composition, or relative species abundances (Terborgh et al. 1996; Yu et al. 1998; Pitman et al. 2001; Condit et al. 2002).

We agree with Hubbell that the classical niche paradigm, which was often based on phenomenological (e.g., Lotka-Volterra) models of interspecific competition, needs to be rethought from first principles. We do not, however, agree that the solution lies with neutral models. In this book, we attempt to resurrect the concept of niche. That is, we heed Hubbell's suggestion that the niche concept needs to be refined. In doing so, we do not develop entirely new models and ideas, but rather link previously unconnected ideas into a single common framework of both classical and contemporary ideas. In chapters 6 and 11, we return to comparisons between the niche and neutral approaches and suggest avenues where the process of "ecological drift," which arises from the connections among local communities in a region (i.e., a metacommunity), can be fully integrated into a niche-based framework.

In this book we seek to accomplish several things. First, we develop a framework around the niche concept that better accommodates several recent insights about niche relations in ecology. Second, we use this updated niche framework to point the way to new questions, conclusions, and syntheses. Finally, we link our interpretation to some of the insights developed using alternative approaches and identify those areas where the greatest challenges for the field of ecology in relation to the niche concept lie.

1.2. History, Evolution, and Divergence of the Niche Concept in Modern Ecology

We start with a brief history of the ecological niche concept. We broadly focus on the historical background, ignoring many of the finer subtleties of usage and precedence that might be of greater historical interest. The history of the niche in ecology and evolutionary biology has been already reviewed by a number of authors (e.g., Hutchinson 1957, 1978; Udvardy 1959; Whittaker et al. 1973; Vandermeer 1972; Colwell and Fuentes 1975; Whittaker and Levin 1975; Giller 1984; Schoener 1989; Griesemer 1992; Colwell 1992; Leibold 1995).

The word *niche* (also *nitch*) has been used in the English language for hundreds of years (see Schoener 1989 for some early uses of the term) and has several related meanings, most of which are not associated with biology at all. The fifth edition of *Merriam-Webster's Collegiate Dictionary* (1959) defines niche as: "1) a recess in a wall, 2) a covert or retreat, 3) a place, condition of life or employment or the like, suitable for the capabilities or merits of a person or qualities of a thing." The third definition is closest to how biologists view a species' ecological niche. However, in the tenth edition of *Merriam-Webster's Collegiate* Dictionary (2001) the first definitions remain, but there are two added ecological definitions: fourth, "a habitat supplying the factors necessary for the existence of an organism or species"; and fifth, "the ecological role of an organism in a community esp. in regard to food consumption." It is interesting that this nonscientific dictionary has two distinct definitions of niche related to ecology and evolutionary biology: one referring to the habitat a species needs for survival, and the other referring to a species' role in or impact on the community. In fact, these definitions, although they seem superficially similar, are indicative of a similar dichotomy among ecological definitions of the niche. Specifically, Grinnell's (1917) original description of the niche, as well as Hutchinson's (1957) well-known definition of the niche as an "*n*-

dimensional hypervolume," are aligned with the first ecological definition, that of ecological requirements. Alternatively, Elton's (1927) usage, as well as MacArthur and Levins's (1967), was more closely aligned with the second, pertaining to ecological roles (see also Schoener 1989; Colwell 1992; Leibold 1995). These distinct definitions have historically caused confusion among ecologists, and as we discuss below in this chapter and more explicitly in chapter 2, by considering both effects in concert we can gain a more complete view of the niche.

The definition of a species' niche as the habitat or role of a species obviously has roots that predate the modern origin of the concept. Genesis clearly states that each species fills a specific role (see also Patrick 1983). Thus, when Noah was charged to build an ark to save all sorts of species from their peril in the Great Flood, the direct implication was that each species allowed on the ship filled a different role that was needed (at least by humans if not by their ecosystems) in the postdiluvian world. Other religions are also founded on writings that refer to species of plants and animals filling various roles within the environment. More explicitly, the writings of early philosophers such as Aristotle and naturalists such as Linnaeus (1758) obviously refer to concepts of diversity and the ecological niche by defining and describing differences among species' traits.

Darwin's (1859) and Wallace's (1876) writings on natural selection and evolution also implicitly refer to species' roles in much the same context that we use the niche concept today. For example, as cited in Stauffer (1975), Darwin used the term "line of life" (in much the same way as we would use "line of work" about a person's job) to describe the role of a species and explain the great diversity of butterflies discovered by Bates in the Amazon basin. Darwin's "line of life" was used similarly to the way a modern biologist would use "niche."

Following the pioneering works of Darwin and Wallace, several naturalists discussed concepts related to a species' niche (see citations in Hutchinson 1978), but Johnson (1910), in a rather obscure publication on the habits and habitats of ladybird beetles, was the first to use "niche" in an explicitly biological way. Despite Johnson's chronological precedence, however, it is widely accepted that the real founder of the niche concept was Grinnell, who in a series of papers, discussed the niches of a variety of species, including their abiotic requirements, habitat, food, and natural enemy relationships (Grinnell 1914, 1917, 1924; Grinnell and Swarth 1913). Although Grinnell's 1904 paper did not use "niche," it was an earlier formulation of the concept to which he later applied the word.

Grinnell's papers, most notably the now classic "The niche-relationships of the California thrasher" (1917), discussed the niche as the place in an environment that a species occupies. Grinnell used the concept to verbally map all of the necessary conditions for a species' existence, including physiological tolerances, morphological limitations, feeding habits, and interactions with other members of the community (notably predators). Grinnell also alluded to some topics of ecological investigation that are not generally accredited to him (Schoener 1989). For example, Grinnell (1904) expanded on the observations of Darwin and Wallace and pioneered the concept that two species must differ in some traits related to their fitness in order to coexist. This "competitive exclusion principle" (Hardin 1960) is generally credited to Gause (1936; see below). By the time he wrote his 1917 paper, Grinnell had discussed this principle as "axiomatic." Grinnell also discussed whether communities were full or whether they were undersaturated, with "empty niches." Many of these questions remain at the forefront of ecological research today.

Elton (1927) provided the second major advance in the use of niche. Specifically, Elton focused on the niche of a species as its functional role within the food chain ("food cycle" in Elton's terminology) and its impact on the environment (e.g., what it eats). The distinction between Elton's focus on the effects of a species on the environment and Grinnell's focus on the effects of the environment on the species will play an important role as we develop our refined niche concept in chapter 2 (see also Schoener 1989; Colwell 1992; Leibold 1995). Many authors, having found little evidence that Elton was aware of Grinnell's writings, have suggested that Elton's conceptualization of the niche was more or less independent of any influence from Grinnell (Hutchinson 1978; Schoener 1989; Griesemer 1992).

Around the time that Grinnell and Elton were refining their niche concepts in animal community ecology, plant ecologists were formalizing their own ideas and pioneering many similar concepts regarding niche and community structure (but often using different terminology). However, as pointed out by Jackson (1981), many of the concepts from early plant ecologists have been subsequently ignored by the vast majority of modern ecologists. Tansley and Salisbury introduced a variety of concepts that are alarmingly similar to ideas that have generally been attributed to authors decades later. For example, Tansley (1917) performed experiments that showed how plant species competed and coexisted, in a sense vying for shared niche space. Tansley also explicitly contrasted the conditions in which a species could theoretically exist with the actual conditions in which it did exist: ideas generally attributed to Hutchinson (1957) in his discussion of "fundamental" and "realized" niches (see below). Salisbury (1929) furthered this distinction and suggested that the similarity in species requirements was strongly related to the intensity of their competition—much the same concept as appears in the more widely appreciated work of Gause (1936).

Although the term "niche" was not a component of the pioneering theories of competition and predation by Volterra (e.g., 1926) and Lotka (e.g., 1924), the elements of these models, depicting interacting populations as dynamical differential equations, provided much of the foundation for modern explorations into the niches of organisms. This was especially evident during the heyday of "niche theory" pioneered by Robert MacArthur and his colleagues, which we discuss below. Gause (1936) made elegant use of experimental laboratory protist populations to explicitly simulate the dynamics of the Lotka-Volterra models. Gause's experiments are perhaps best known for their part in demonstrating the principle of competitive exclusion, sometimes referred to as the "Volterra-Gause principle" (Hutchinson 1978), which states that no two species with identical competitive effects on one another can coexist locally. Although this principle was evident in Darwin and Wallace's writings and was formalized by Grinnell, it was not until Gause's strong experimental approach that this axiom became widely known. In fact, competitive exclusion remains one of the most fundamental principles in ecology today (e.g., Chesson 1991, 2000b; but see Hubbell 2001 for arguments to the contrary).

Following these early uses and refinements of the niche concept in both animal and plant ecology, Hutchinson provided a major (sometimes called "revolutionary"; see Schoener 1989) step in the definition and quantification of the niche concept. Hutchinson's definition was first presented in a footnote of a limnological paper (Hutchinson 1944, 20 n. 5). Hutchinson stated, "The term niche (in Gause's sense, rather than Elton's) is here defined as the sum of all the environmental factors acting on the organism; the niche thus defined is a region of an n-dimensional hyper-space." Hutchinson's (1944) "n-dimensional hyper-space" transformed to an "n-dimensional hypervolume" in his now classic concluding remarks at the Cold Springs Harbor Symposium of Quantitative Biology (Hutchinson 1957). In this paper, Hutchinson applied a more quantitative approach to the niche concept than had been done previously by Grinnell and Elton. To do this, Hutchinson explicitly defined any number (n) of limiting factors (e.g., temperature, resources) for a given organism. The quantity of each limiting factor a given organism needs to exist can then be plotted in a hypervolume

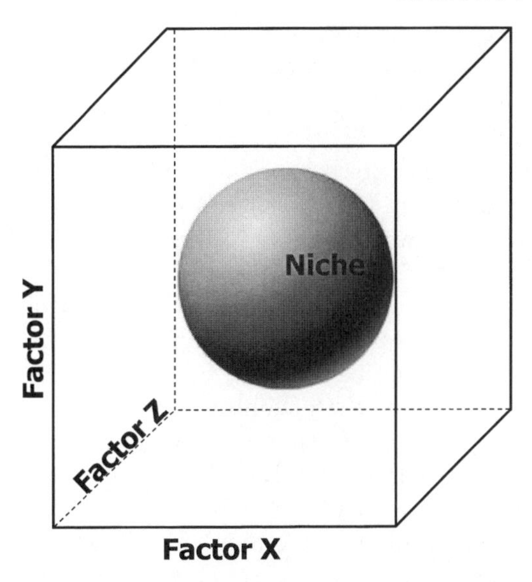

Fig. 1.2. Hypothetical depiction of a three-dimensional volume (three factors) within Hutchison's (1957) *n*-dimensional hypervolume niche. The area within the cube represents the total available amount of each factor, while the area within the sphere represents the amount of each factor needed for a given species to survive; i.e., its niche.

or space of n dimensions. The space occupied within the n-dimensional hypervolume would thus be the range of conditions where a species could exist (see also Hutchinson 1965, 1978) (fig. 1.2). The n-dimensional hypervolume approach provided a conceptually quantitative niche concept, which provided a useful way to succinctly define a niche, as well as to terrify math-phobic introductory ecology students (this definition remains the mainstay of modern ecology textbooks: e.g., Pianka 1995; Begon et al. 1996; Krebs 2000). Around the same time, MacFadyen (1957) presented a similar, but less well developed, description of the niche in multiple dimensions.

Hutchinson (1957) took his quantitative formulation of the niche a step further. As both animal and plant ecologists had long realized, the areas and conditions in which a species could feasibly live were often greater than those where the organism actually lived, and this was typically caused by the effects of interactions among co-occurring species. Thus, Hutchinson termed a species' *fundamental niche* as all aspects of the n-dimensional hypervolume in the *absence* of other species, and the *realized niche* as the part of the fundamental niche to which the species was restricted due to interspecific interactions (we return to this distinction in chapter 3). In so doing, Hutchinson revolutionized the niche concept from the vague constructs of Grinnell, Elton, and others into a quantifiable unit that allowed explicit theoretical analyses and prediction (e.g., Levins 1968; MacArthur 1969; Vandermeer 1972; Schoener 1989).

Hutchinson's other major contribution to the niche concept and its role in understanding patterns of species diversity came in his 1959 address to the American Society of Naturalists entitled "Homage to Santa Rosalia, or why are there so many kinds of animals?" (Hutchinson 1959). In this address, Hutchinson weaves a wonderful story, beginning with a visit to a small artificial pond below a cave thought to house the remains of the Italian saint Santa Rosalia. In this pond, as throughout much of Europe, Hutchinson noted the coexistence of two abundant species of Corixidae (Hemiptera) insects (commonly referred to as water boatmen) (fig. 1.3). In reflecting upon why there were only two common species in these ponds, Hutchinson began to wonder about the causes and limits of the number of species we see. Giving rise to what is now a blossoming area of research on determinants and consequences of biological diversity (e.g., Ricklefs and Schluter 1993; Rosenzweig 1995; Gaston 2000; Hubbell 2001), Hutchinson discussed several factors that might either cause or limit the amazing diversity of plants and animals in nature. Armed with his own niche concept (Hutchinson 1957), he set out to determine what limited the similarity of coexisting species. In looking at a variety of size ratios, he found what he termed a tentative relationship—when two similar species coexist, the mean ratio of the size of the larger to the smaller is 1.3:1. This soon became known as the Hutchinsonian ratio (Lewin 1983) and consumed much of the creative interests of both theoretical and empirical ecologists for many years (Karieva 1997).

MacArthur and his collaborators greatly expanded Hutchinson's approach and motivated an extraordinary amount of creative energy by ecologists in the 1960s and 1970s. This led to a large body of work

Fig. 1.3. Drawing of two species of Corixidae (Hemiptera) water boatmen similar to those that inspired Hutchinson's (1959) ideas about species diversity. Note that the two species differ in size by a ratio of approximately 1.3:1. Drawing courtesy of Lynn Adler.

on what is now known as niche theory, and the concept became firmly ensconced in most problems of ecological study. Niche theory was essentially a group of theoretical models designed to investigate how many and how similar coexisting species could be within a given community (MacArthur 1969, 1972; MacArthur and Levins 1967; May and MacArthur 1972; for a review, see Vandermeer 1972). They were invariably based on the Lotka-Volterra equations and were coupled with the widespread view that interspecific competition was very important in structuring natural communities. This theoretical framework drove field ecologists to measure niche breadth (i.e., the variety of resources or habitats used by a given species), niche partitioning (the degree of differential resource use by coexisting species), niche overlap (the overlap of resource use by different species), and niche assembly (the colonization and organization of species with different resource use in new or abandoned habitats) in natural communities (see e.g., MacArthur 1969; Schoener 1968, 1974a; Colwell and Futuyma 1971; Roughgarden 1972; Pianka 1973; Diamond 1975). Niche theory consequently became the focus of a number of studies on topics from the realms of evolutionary ecology, population and community ecology, and biogeography.

Hutchinson, MacArthur, and others used the idea of competition for resources as the primary underlying mechanism driving ecology. Although both Hutchinson and MacArthur also considered many other factors, such as predation and environmental variability, subsequent authors focused on their work on resource competition. The word "niche" became firmly entangled with the notion of interspecific competition (fig. 1.4). Prior to 1965 less than 50 percent of all articles that contained the word "niche" also had "competition" (though during the ten-year period 1936–45, this association did exceed 50 percent but then declined), suggesting that authors used "niche" quite often without considering competition. However, by 1975 this figure increased to around 80 percent, where it has remained. We explored the connection between niche and two other common species interactions, predation and mutualism, and found no similar strong correlations. Given the strong linkage between niche and competition, it is not surprising that as resource competition has fallen out of favor as the overriding factor influencing community structure, so has the niche concept.

1.3. The Downfall of the Niche Concept

The ardent way that many ecologists approached niche and competition theory brought about a revolution in the way that ecological ques-

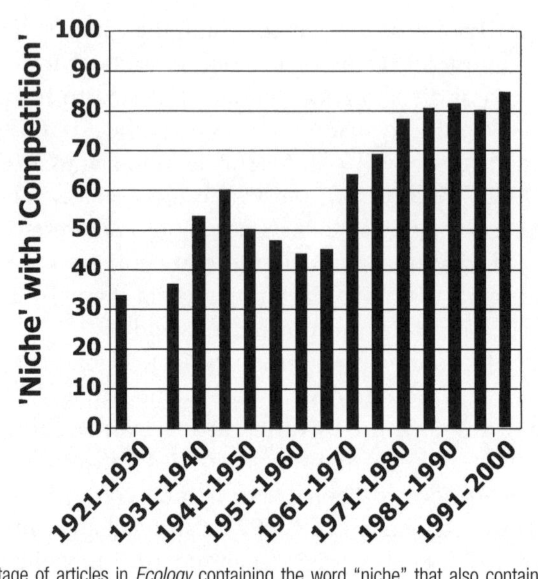

Fig. 1.4. Percentage of articles in *Ecology* containing the word "niche" that also contained the word "competition." Data were obtained as in fig. 1.1.

tions are asked and answered. The backlash that resulted from the boom of niche theory in the late 1970s was couched in terms of the philosophy of science, in particular the hypothetico-deductive methodology espoused by Popper (e.g., 1963), which emphasizes the comparison of null hypotheses (i.e., the result expected under random conditions) with the alternative hypotheses of interest (i.e., the result due to some condition of interest). Simberloff, Strong, and their colleagues rightly pointed out that the explosion of studies exploring patterns of competition and niche theory were typically without adequate null hypotheses (e.g., Simberloff 1978; Connor and Simberloff 1978, 1979; Strong et al. 1979; Simberloff and Boecklen 1981). Thus, the validity of scores of studies examining patterns of character displacement, coexistence and diversity, and biogeography based on competition and niche theory were brought into question. Lewin reviewed the professional and personal acrimony that surrounded this debate in a provocative article, "Santa Rosalia was a goat" (1983), the title referring to the discovery that the bones thought to be those of Santa Rosalia, whom Hutchinson (1959) dubbed the patron saint of diversity studies, were actually those of a goat (see Simberloff and Boecklen 1981).

Most researchers eventually conceded that appropriate null hypotheses and experimentation were necessary in order to disclose the importance of competition and niche theory (see, e.g., Gotelli and Graves 1996). However, the debate about the form of null models remains contentious (see, e.g., Losos et al. 1989; Brown et al. 2000; Stone et al. 2000). A particularly caustic exchange between the groups can be seen in the discussion between Connor and Simberloff (1984) and Gilpin and Diamond (1984), where the line between scientific discourse and name calling became blurred. The backlash created by this exchange arguably changed the face of ecology (e.g., Lewin 1983; Colwell and Winkler 1984; Harvey et al. 1983; Gotelli and Graves 1996; Brown 1997). Among other things, the ubiquity of resource competition has been downplayed for a more pluralistic view that includes the role of other factors, such as predation and abiotic stresses. Furthermore, hypothesis testing and rigorous statistics, as well as a strong emphasis on experimental approaches, have emerged to the forefront of ecology. Despite this positive response, the emphasis on experimental and statistical rigor necessarily created a focus on smaller-scale, shorter-term processes that were amenable to such manipulations. Studies of large-scale diversity, range, abundance, and the like were diminished, while studies of local interactions and processes were emphasized (see, e.g., Brown 1995; Maurer 1999).

Even before traditional tests of niche theory were being criticized in the empirical literature for their lack of statistical rigor and null models, and experiments were pushed as the primary way to discern the importance of competititive interactions and coexistence, theoreticians were refining competitive models to more explicitly incorporate the mechanisms of species interactions. MacArthur (1972) showed that the competition coefficient and carrying capacity of the Lotka-Volterra equations could be linked with more mechanistic consumer-resource models under certain limiting assumptions. Others, including Schoener (1974b) and Abrams (1975, 1983), more specifically discussed niche overlap and the limiting similarity of coexisting species in the context of the linkage between the Lotka-Volterra models and more mechanistic approaches. Although in the limiting case, Lotka-Volterra models and mechanistic consumer resource models converge to similar structure (see also Tilman 1982; Petraitis 1989), theoreticians also began to explore the limitations of the earlier niche theory.

From this, a diverse body of theoretical literature arose that discussed several aspects of species niches and coexistence as it related to competitive interactions (e.g., Abrams 1977, 1980, 1986), the influence of predation (e.g., Holt 1977, 1984, 1985, 1987), and intrinsic and extrinsic spatial and temporal heterogeneity (e.g., Armstrong and McGehee 1976, 1980; Chesson and Warner 1981; Chesson 1985; Warner and Chesson 1985). However, these advances in theory were not well connected with much of the empirical focus. This was most likely the result of two related factors. First, empiricists were focused on testing very basic hypotheses about the presence or absence of the effects of interspecific interactions (mostly competition) in experiments or against statistically rigorous null models. Second, empiricists were skeptical of the utility of theory in the aftermath of the hostile reaction to niche theory discussed above.

A reasonable caricature of how science progresses can be seen in the *dialectical* approach (e.g., Levins and Lewontin 1980). The dialectical cycle begins with a thesis, followed by skepticism or antithesis, and finally synthesis. We view the niche concept as having progressed in a dialectical manner (fig. 1.5). Beginning with the formulations of Grinnell and Elton, the niche concept became a full-blown thesis with Hutchinson and MacArthur. The flaws of this thesis were pointed out by the antithesis led by Simberloff and Strong, who called for more rigorous null models and experimentation. Following this antithesis, empirical ecologists shifted their focus using strict philosophical and statistical ideals and well-controlled experimentation. However, in the backlash of this antithesis, many questions of broad ecological interest related to the niche concept have fallen to the wayside.

1.4. Revisioning the Niche Concept

The history of the niche concept draws on several important issues that must be resolved in order for it to have a useful, synthetic role in ecology. First, it needs to incorporate processes other than resource competition. While previous concepts of the niche have often had this broader perspective, today the niche concept is often too closely aligned with competition even though theoretical and empirical ecologists have embraced other processes. Second, the concept needs to clarify the distinction between things that describe how an organism responds to the environment (implicit in concepts focused on requirements such as Grinnell's and Hutchinson's) and things that describe how an organism alters its environment (implicit in concepts focused on impacts such as Elton's and MacArthur's). Finally, the idea of the niche needs to be made relevant to multiple spatial scales. Current niche theory is often too narrowly focused on explaining species interactions at the local scale (where population dynamics are the only processes present), but many of the more important and challenging eco-

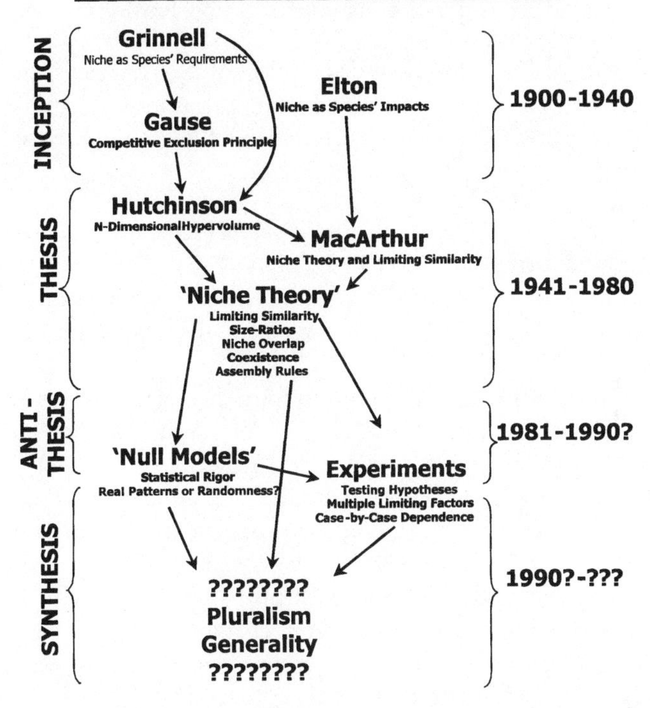

Fig. 1.5. The progression of the niche concept through time, with the associated dialectical progression of ideas.

logical questions occur at larger scales (where colonization dynamics, invasions, and the like become quite important).

In order to put the niche concept into the context of a synthesis for ecology, we propose a revised definition. First, we give a broad definition:

NICHE DEFINITION #1: the joint description of the environmental conditions that allow a species to satisfy its minimum requirements so that the birth rate of a local population is equal to or greater than its death rate along with the set of per capita effects of that species on these environmental conditions.

This definition is a simple joining of the two concepts that we have outlined in our historical review. However, we also want to use this definition to develop a tool that can flexibly address and help synthesize the wide variety of ecological phenomena under current study. To do this we will use simple population dynamics models (in chapter 2) to justify a second more precise version of this definition as:

NICHE DEFINITION #2: the joint description of the zero net growth isocline (ZNGI) of an organism along with the impact vectors on that ZNGI in the multivariate space defined by the set of environmental factors that are present.

This definition is laden with mathematical jargon that corresponds to the first definition but is key to letting us use a number of analytical and graphical methods that can address complex interactions in ecology. It also makes the general concepts in the first definition more precise.

We develop this second definition in the next chapter drawing heavily on work by MacArthur (1972), Tilman (1982, 1988), Holt et al. (1994), Grover (1994, 1995, 1997), Wootton (1998), Leibold (1995, 1996, 1998), and Chase, Leibold, and Simms (2000). Tilman's work in particular is found in numerous basic texts (Begon et al. 1996; Krebs 2000; Ricklefs and Miller 2000). Readers who are particularly familiar with this body of theory will probably not lose much of our basic message by skimming through chapter 2. We then proceed throughout the rest of the book to use our revised definition (especially using the more precise second version) to evaluate a diverse array of ecological problems.

1.5 Context for This Book

Following the overemphasis on simple, single-hypothesis factorial experiments that characterized many empirical ecological studies of the 1980s and 1990s, contemporary ecology has become much more synthetic. Some of the best studies today combine observational and pattern analyses, theoretical predictions, and experimental approaches. This pluralism has allowed a much deeper understanding of pattern and process that is not possible with a singular approach. Contemporary ecology has become of utmost applied importance as a result of several environmental crises, including pollution, habitat degradation, species extinctions, and the looming compounding dangers foreseen in the exploding human population (Quammen 1996; Reaka-Kudla et al. 1997; Pickett et al. 1997; Levin 1999; Balvanera et al. 2001; Daily et al. 2001; Balmford et al. 2002). The importance of these environmental questions and syntheses is, in part, forcing current ecology to again focus on large-scale questions including: species resource utilizations

(MacNally 1995), patterns and processes causing species diversity on local and regional levels (e.g., Ricklefs and Schluter 1993; Rosenzweig 1995; Huston 1999; Amarasekare 2000; Gaston 2000; Bell 2001; Hubbell 2001; Mouquet and Loreau 2002), effects of species diversity on ecosystem process and function (Tilman, Wedin, and Knops 1996; Tilman, Lehman, and Thompson, 1997; Tilman, Knops, et al. 1997; Naeem 1998; Tilman 1999; Loreau 2000a; Naeem et al. 2000; Tilman, Reich, et al. 2001; Kinzig et al. 2002), effects of regional species turnover on community interactions (Leibold et al. 1997), species distribution and abundance (e.g., Brown 1995; Maurer 1999; Gaston and Blackburn 2000), community invasibility (Williamson 1996; Levine and D'Antonio 1999; Shurin 2000; Shea and Chesson 2002), large-scale spatial and heterogeneous processes (Pacala and Tilman 1994; Tilman and Karieva 1997; Chesson 2000a,b; Dieckmann et al. 2000), and speciation and extinction's effects on biodiversity and distributions (Losos and Schluter 2000; Schluter 2000, 2001; Hubbell 2001; Godfray and Lawton 2001).

In many ways, these questions are very similar to many of those addressed by Hutchinson, MacArthur, and others over thirty years ago. However, while Hutchinson and MacArthur focused on the niche concept as a foundation for their studies, today the word "niche" is often replaced with other terminology in order to disassociate from the entanglement between niche and competition. Terms such as "versatility" and "trade-offs" have replaced "niche breadth" and "niche divergence." Nevertheless, the niche concept as conceived by Grinnell and Elton and revolutionized by Hutchinson and MacArthur still underlies most ecological questions (e.g., Chesson 1991; Leibold 1995; Pianka 1995; Tilman 1999). We believe the change in terminology is largely cosmetic and that when closely examined all of these approaches are really focused on understanding the niche relations among species and their environment.

If current usage of the term "niche" is so confusing and ambiguous, is it worth our while to salvage and revise it? One prominent ecologist suggested that we should "allow the term to die a natural death" and replace it with other analogous or more specific terms. Furthermore, Hubbell (2001) claims that the niche concept is not needed at all to understand fundamental patterns in ecology. However, we feel that the niche has provided and can continue to provide the central conceptual foundation for ecological studies. In addition, we worry that as the use of niche terminology in ecology declines, so does the understanding of the importance of much of the earlier work done in ecology (see also Graham and Dayton 2002). This has caused substantial reinventing of the wheel. What we hope to accomplish in this book is to recast the niche concept in a more modern and synthetic way.

1.6. Summary

1) The niche concept is important both as a tool for thinking about ecological and evolutionary phenomena and as a synthetic device for integrating these phenomena across levels of organization (e.g., individuals to ecosystems).
2) Recently, use of the niche concept has precipitously declined. Furthermore, recent "neutral theories" (e.g., Hubbell 2001) claim that the niche concept is not necessary for understanding broad-scale ecological patterns.
3) There is tremendous confusion about the concept because previous authors have not consistently distinguished between the responses of organisms to their environment and the effects of organisms on their environment.
4) Due to its tight connection with interspecific competition, the use of the niche concept precipitously declined as the perceived importance of competition declined, and other processes were found to have important roles. The recognized need for rigorous null hypotheses, statistics, and experimentation caused subsequent ecological investigations to focus on local-scale questions and deemphasize larger-scale phenomena.
5) Modern investigations have again begun to focus on larger-scale processes, but are only loosely connected to the highly relevant earlier studies. We propose that the niche concept, in revised form, can provide the synthetic tool to link classical and modern ideas and approaches.
6) We provide a revised definition of the niche concept focused on recognizing the importance both of a species' requirements and its impacts on the ecosystem.

CHAPTER TWO

REVISING THE NICHE CONCEPT: DEFINITIONS AND MECHANISTIC MODELS

In chapter 1, we discussed distinctions among the ways previous authors have defined the concept of a species' niche. How can we develop a niche concept that brings these ideas together into a single organized conceptual framework? We propose that a species' niche can be divided into two main components. Grinnell (1917), Hutchinson (1957), and others focused on a species' *requirements* for survival in a given environment. On the other hand, Elton (1927), MacArthur and Levins (1967), and others focused on the *impacts* of a species on its environment. Aspects of both of these components are relevant for different conclusions that one might make about niche relations and their implications for species interactions.

Throughout this book, we will define a species' niche quantitatively in the context of its requirements and impacts using both analytical and graphical approaches. While we will emphasize the graphical models because we feel that they allow a more intuitive understanding of the predictions, we emphasize that the models are based on a solid analytical framework. In this chapter, we will develop the synthetic niche concept that we will use in the rest of the book, and we will present examples of the very simplest cases. We define the *niche* of a species as *the environmental conditions that allow a species to satisfy its minimum requirements so that the birth rate of a local population is equal to or greater than its death rate along with the set of per capita impacts of that species on these environmental conditions.*

The graphical/analytical framework that we use builds upon MacArthur's (1972) consumer-resource models, which were extended and popularized by Tilman (e.g., 1980,1982) and are now common in a wide variety of contemporary modeling studies (e.g., Holt et al. 1994; Grover 1994, 1995,

1997; Grover and Holt 1998; Leibold 1995, 1996, 1998; McPeek 1996; Chase 1999a; Chase, Leibold, and Simms 2000). Justification and proofs for this analytical framework are presented in detail by MacArthur (1972), Tilman (1980, 1982), Holt et al. (1994), Leibold (1995, 1996, 1998), and Grover (1997). We downplay the analytical models because they have been discussed at length elsewhere and because they are not necessary in order to understand the fundamental predictions of the models. However, for completeness, we present some of the simplest analytical results in a box and in the appendix to this chapter. Readers familiar with these relationships can probably skip these, and those familiar with the basics of these graphical models can probably skim over much of the material in this chapter.

2.1. Interactions Involving a Single Niche Factor

We begin by describing the interaction between a single organism and a single niche factor. This interaction can be separated into two parts. We call the first the *requirement component*—how the magnitude of the factor determines the fitness of the organism. The second is the per capita effect of the organism on that factor, which we call the *impact component*. We start by looking only at factors with *direct* effects. We will then use this information to look at interactions mediated by *indirect* effects (sensu Miller and Kerfoot 1987; Wootton 1994), species interactions, and species coexistence.

The simplest way to depict the responses of a species to a particular factor is to examine how its birth and death rates vary along a gradient of that factor. The most common factor that has been examined in this way is a consumable resource, such as food or nutrients (fig. 2.1a). In this case, we assume that birth rate is an increasing function of resource availability and that death rate is independent of resource availability (but other functional forms could easily be substituted). Of course, the response of an organism to resources is not just hypothetical. For example, laboratory studies on bacteria (e.g., Hansen and Hubbell 1980; Bohannan and Lenski 1997), algae (Tilman et al. 1976; Grover 1988, 1991), and planktonic rotifers (Rothaupt 1988) show predictable growth rate functions in response to increasing food, and often these growth functions are not linear, but saturating (see review in Grover 1997). Similarly in the field, Tilman and Wedin (Tilman and Wedin 1991a,b; Wedin and Tilman 1993) showed that the growth rates of several plant species increased with resources. Other field experiments showed that the growth rates of animal species increased with resources in both aquatic (e.g., Leibold 1989; Osenberg 1989;

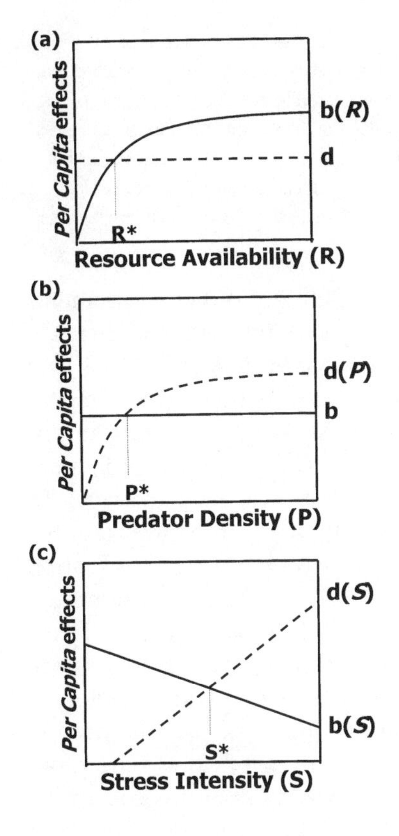

Fig. 2.1. Examples of the niche requirement of a species for a factor. In each case we draw the birth rate *(b)* (solid line) and death rate *(d)* (dashed line) along a gradient of that factor. (a) Requirements of a species for a consumable resource *(R)*. In this case, we assume that the birth rate depends on resource availability *(b(R))*, and that the death rate is constant. The point at which the birth rate crosses the death rate is the equilibrium level of resources (R*) where the consumer population maintains zero net growth ($dN/dt = 0$). When resource levels are above this point, the population will grow, whereas when they are below this point, the population will decline. (b) Requirements of a species with a predator (P). Here, we have assumed that the birth rate is constant, but the death rate varies negatively as a function of *P (d(P))*. The intersection of the birth and death functions indicates the density of predators in which the population can maintain zero net growth, above which the population will decline and below which the population will grow. (c) Requirements of a species to stress (S). In this case, we assume that birth rates are a negative function of the intensity of stress *(b(S))*, and death rates are a positive function of the intensity of stress *(d(S))*. The intersection *(S*)* represents the level of stress where the population has zero net growth, above which the population will decline and below which the population will grow.

Chase 1999b) and terrestrial (e.g., Moen et al. 1996; Schmitz 1997) systems.

We can use methodology similar to that used to examine consumable resources to consider how a species' birth and death rates (requirements) respond to any factor that influences them, such as predators or stress agents. For example, in the case of predators, the direct interaction will be negative, since predators increase the death rate (d is a function of predation [$d(P)$]), as in fig. 2.1b. Predators could, however, also decrease the birth rate in cases where prey behaviorally or morphologically respond to the presence of predators. The hypothetical stress agent we depict in fig. 2.1c imposes a linear increase on the organism's death rate ($d(S)$) as well as a linear decrease on its birth rate ($b(S)$). Some stresses, such as periodic disturbances (e.g., fire and floods), may affect only d and not b, and yet other stresses may have more complicated effects, such as changing resource uptakes and requirements.

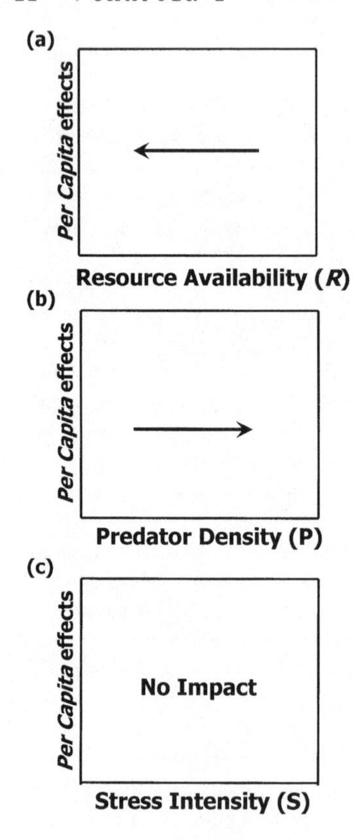

Fig. 2.2. Examples of the impact a species can have on a factor. In each case, the length of the impact is proportional to the instantaneous per capita impact of the species on the factor. (a) Impact vectors of a consumer on its resource (negative). (b) Impact vectors of a prey on a predator (positive). (c) No effect of species on an abiotic stress.

In addition to the requirements of a species for a particular factor, which correspond to Grinnell and Hutchinson's usage of "niche," a species will often also have an impact on that factor, which is more consistent with Elton's usage. To quantify this effect, we use the term *impact vector* as a generalization of Tilman's (1982) "consumption vector." If, for example, the factor is a resource, the organism will tend to lower its concentration or density (fig. 2.2a), and the impact vector will point toward the *y*-axis of decreasing resource concentrations. If the factor is a predator (fig. 2.2b), the organism will tend to increase its density, and the impact vector will point away from the *y*-axis toward increasing densities of predators. Finally, if the factor is a stress agent, the organism will generally have no impact (fig. 2.2c). These interactions can be described using vectors as in fig 2.2 to indicate the direction and the magnitude of the interaction. The length of the vector is proportional to the instantaneous per capita effect size (or impact) (for an explanation of the implications of such effect sizes, see, e.g., Laska and Wootton 1998; Berlow et al. 1999; Abrams 2001a).

With a single factor, we have discussed the two aspects of an organism's niche quantitatively—its requirements and its impacts. Although the salient points of this model do not require a specific analytical solution, we derive a model of requirements and impacts using the basic consumer-resource formalism in the boxed text for those who prefer thinking about these relationships analytically rather than graphically.

We can also use this model to make a simple but fundamental point about species coexistence. In the cases described above, we have fo-

Deriving the Resource Requirements and Impacts of an Organism

We use a very simple model with linear effects and no population structure (note that this model is slightly different from fig. 2.1.a, which assumes an asymptotic relationship; this change in assumptions does not influence the basic conclusion).

Let $dN/dt = N(faR - d)$
$dR/dt = c(S - R) - fNR,$

where N is the density of the consumer, R is the density of the resource, f is the per capita rate that the consumer encounters and consumes, a is the efficiency with which the consumer converts resources into new consumers, and d is the density-independent death rate of the consumer. We further assume that the resource renews at rate c from the resource supply (S) that is not already bound up in R; thus, the total renewal rate of the resource is $c(S - R)$. From these equations, the consumer's minimum resource requirement at equilibrium is $R^* = d/fa$. We can also calculate the consumers' per capita impact on the resource as fR. When the total impact of the consumer population on the resource is equal to the renewal rate of the resource, then the resource levels will be at an equilibrium. Here, $fNR = c(S - R)$ and $N = c(S - R) / (fR)$. From this, we can substitute resource density at equilibrium, R^*, for R and solve, which gives $N^* = ac[(S - d) / (fa)]/d$, and the per capita impact of the consumer on the resource at equilibrium as $fR^* = d/a$.

In words then we can define the two components of the niche for a single factor using this formalism. First, R^* is the minimum level of resources required for birth rates to balance death rates, so that the requirement component of the niche will be where $S > d/(fa)$ (resource supply exceeds R^*). Second, at this equilibrium point, the impact of the consumer on the resource is d/a.

We do not similarly derive the equations for predators, stresses, or other factors; some of these are discussed in detail elsewhere (Holt et al. 1994; Wootton 1998). Nevertheless, the basic formalism holds regardless of which niche factor is being considered.

cused on the conditions that just allow a species to renew itself, so that its average birth rate is at least equal to its average death rate. When a population is not growing or declining (i.e., it is at equilibrium), the birth rate is equal to the death rate ($b = d$, where the lines intersect), and this defines the lower boundary of the requirement component of the organism's niche factor. This point also corresponds directly to Hutchinson's (1957) definition of a species' niche: the minimum level of a factor at which a species can maintain its population (see Maguire 1973). Where $b > d$, the population can grow, and where $b < d$, the population goes extinct (i.e., it is not a part of the species' niche).

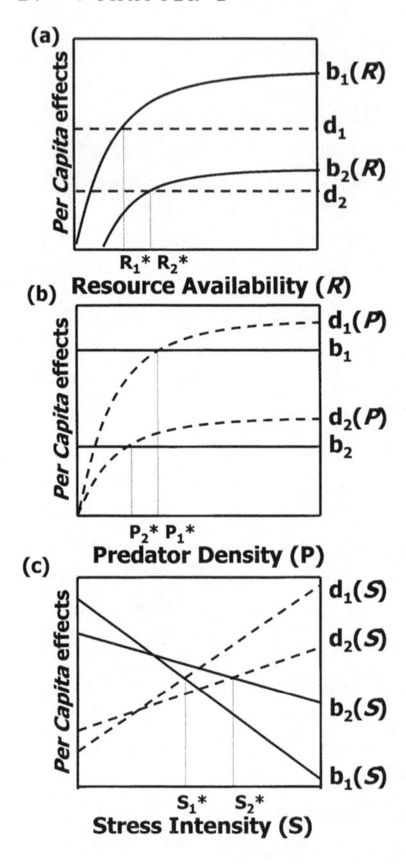

Fig. 2.3. Examples of the interaction between two species that utilize a single niche factor. (a) Two species consuming a single resource. In this case, the species with the lowest R* (where birth rates equal death rates) will outcompete the other and exist alone. (b) Two species consumed by a single predator. In this case, the species with the highest P* will exist alone. (c) Two species influenced by an abiotic stress. In this case, the ultimate winner of the interaction will depend on the intensity of the stress.

When the factor is a resource, the minimum level at which the organism can survive (where $b = d$) is termed R* (e.g., Tilman 1982); when the factor is a predator, the maximum level at which the organism can survive is termed P* (Holt et al. 1994); and when the factor is a stress, the maximum level at which the organism can survive is termed S*.

Armed with these definitions, we can easily show that two species cannot coexist at equilibrium unless there is more than one limiting factor (see below) or there is intrinsic or extrinsic environmental variability through space or time (see chapter 6). For example, if two species compete for a single limiting resource, the species with the lower R* will always outcompete the species with the higher R* and exist alone (Tilman 1982) (fig. 2.3a). This is because both species have impacts on the resource, but the one with the lower R* can reduce and survive on lower resource levels than the other species. That is, the species with the lower R* has birth rates greater than or equal to their death rates on lower levels of resources than does the species with the higher R*. Alternatively, if two species are consumed by a common predator, they will increase the density of predators by feeding them, and the species with the higher P* will exist alone (Holt et al. 1994) (fig. 2.3b). Finally, when two species are both affected by the same stress but have no impact on that stress, the predictions are somewhat more complicated (fig. 2.3c). If the intensity of stress is greater than one species' S^* but not the other's, then the species with the higher S^* will exist alone, whereas if the intensity of stress is greater than either species' S^*, neither will be able to exist. If, however, the intensity of stress is less than either species' S^*, then we cannot predict which species will exist in the absence of further information (such as their abilities to compete for limiting resources).

2.2. Interactions Involving the Interplay between Two Niche Factors

In the preceding section we focused on situations involving only one environmental factor to develop the fundamental definitions of niche requirements and impacts that we will use throughout this book. We did this by assuming that all other environmental factors were constant. Because organisms respond to multiple factors, we next consider the interplay among such factors.

Here we address the question of how to combine two environmental factors to see how they might interact. There are eight qualitatively different situations in which an organism can require and impact a factor (+/−, −/+, −/−, +/+, +/0, −/0, 0/+, and 0/− [excluding the 0/0 case, where an organism neither requires nor impacts a factor]), depending on whether requirements and impacts are positive, negative, or zero (see table 2.1 for some examples). Rather than illustrate all possible combinations, we will focus on a handful of the better-studied combinations, since these will also serve in later discussions. However, it is interesting to note that some combinations have received almost no attention whatsoever. For example, mutualisms have received very little attention using mechanistic models (but see Schwartz and Hoeksema 1998), though we suspect that it would be a relatively straightforward task to derive models similar to those discussed here.

For perhaps obvious (though not necessarily good) reasons, previous work has largely focused on the interactions that involve trophic transfers; that is, the interaction between consumers and resources. These involve the interplay between a species and (1) two limiting resources (MacArthur 1972; Tilman 1982; Grover 1997), (2) a limiting resource and a predator (Holt et al. 1994; Leibold 1996), or (3) two predators (Holt 1977; Holt and Lawton 1993; Leibold 1998). Holt (1997) has viewed these categories of interactions as community "modules" and has suggested that an understanding of these simple modules can enhance our understanding of more complicated food webs. We add (4) a limiting resource and a stress agent (Wootton 1998).

We present each of these modules in concert, rather than focusing on each module separately for one important reason: all of the models

Table 2.1. Summary of the joint responses and impacts of a species to a factor (another organism or an abiotic factor) and the type of resulting interaction.

Factor	*Response*	*Impact*	*Type of Interation*
Organism	+	−	consumer
Organism	−	+	prey
Organism	−	−	interference competitor[1]
Organism	+	+	mutualist
Organism	+	0	commensal
Organism	−	0	ammensal
Abiotic	−/+	0	stressor/enhancer[2]
Abiotic	+/−/0	+/−	bioengineer[3]

[1] Interference competitors directly interfere with each other, in contrast to exploitative competitors that compete for limiting resources (an indirect interaction; see chapter 3).

[2] A stressor or enhancer is an abiotic factor that either benefits or detracts from the population growth of the species in question. A stressor could be a direct mortality factor or could lower birth rates. An enhancer is an abiotic resource.

[3] A bioengineer is a species that has either a positive or negative impact on the abiotic environment (see, e.g., Jones and Lawton 1995).

Note: These are only some of the possible interactions a species can have with a factor.

provide very similar general conclusions. The mathematics and background for each module are presented in the appendix to this chapter.

Requirements

For two limiting resources (which is probably the best-known module) we can draw a contour map of fitness as a function of the two resources. To keep everything simple and linear, we focus on the case where the two resources are substitutable. Substitutable resources have fixed proportional effects on the fitness of a species, and these effects are independent of the density of the resource. Readers familiar with Tilman's work on resource competition (e.g., 1982, 1988) will note that we are focusing on "substitutable" resources rather than the "essential" resources often emphasized by plant ecologists. Substitutable resources have linear additive effects; the responses to essential resources are not additive. We come back to these distinctions in chapter 5. When two resources are considered instead of one, there are many combinations of the two resources in which the birth rate equals the death rate. All such values taken together define a line in two-dimensional space of combinations of the two resources; this line, at which the species has zero net growth, is termed the *zero net growth isocline* (ZNGI) ($dN/dt = 0$) (fig. 2.4a). As with the case of one limiting factor, we can use Hutchinson's (1957) definition to define the niche boundary as the ZNGI, and the range of the niche as any situation where fitness is equal to or greater than zero ($dN/dt \geq 0$) (see also

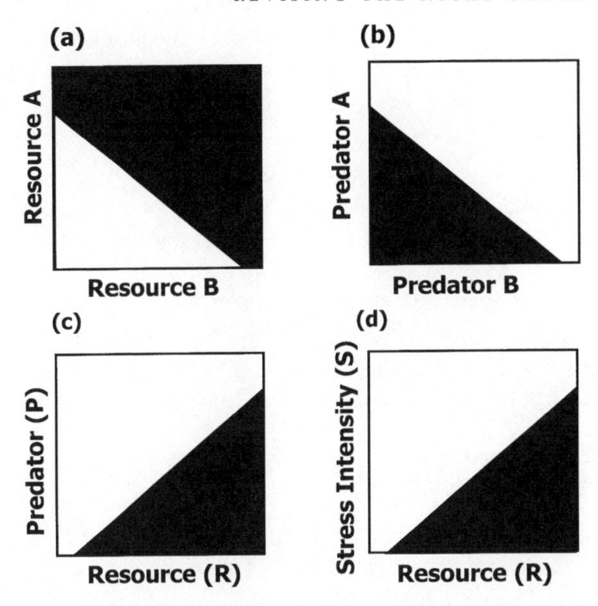

Fig. 2.4. Examples from four of the modules discussed in text of the requirement niche for a species on two factors. The zero net growth isocline (ZNGI), where $dN/dt = 0$ (thick solid line) denotes the boundary of the species' niche (birth rate = death rate). Shaded areas are where birth rates are greater than death rates and are included within the species niche. (a) A species with two substitutable resources, A and B. (b) A species with two predators, A and B. (c) A species with a predator (P) and a resource (R). (d) A stress (S) and a resource (R).

MacArthur 1972; Maguire 1973). That is, the ZNGI defines the requirement component of the niche in two-dimensional space.

In addition to describing the effects of two resources, ZNGIs can be used to describe interactions between any two factors of interest. For example, when a species is consumed by two predators, we draw a ZNGI similar to the one above, except that in this case, the ZNGI represents the maximum amount of predators where the prey can exist, and it will have positive population growth with densities of predators below that amount (fig. 2.4b) (Leibold 1998). Similarly, when species consume a resource and are consumed by a predator, we can view how the predators alter the ability of the species to maintain zero net growth on resources (fig. 2.4c) (Holt et al. 1994; Leibold 1996; Chase, Leibold, and Simms 2000). In the absence of predators, the intercept of the ZNGI on the *x*-axis will simply be that species' R^* for the resource. However, as predator density increases along the *y*-axis, predators will alter death and/or birth rates, and the amount of resources (R^*) that the species will need to maintain zero net growth will increase. The steepness of the slope of the ZNGI indicates how strongly the species is influenced by the predator; the shallower the slope, the greater the effect of predators. Finally, when a species consumes a limiting resource and is influenced by a stress, increases in the level of the stress will increase the amount of resource the species needs in order to maintain zero net growth (fig. 2.4d) (Wootton 1998); the shallower the slope, the greater effect of stresses.

Although we have focused on only a few combinations of limiting factors, it should be possible to consider any combination of factors that positively, negatively, or neutrally influences a species' birth and death rates and to derive a corresponding ZNGI. For example, if a facultative mutualist decreases the resource requirements of a species, we could draw a ZNGI with a negative slope along mutualist and resource axes.

Impacts

The impact of an organism on a single factor can easily be extended to two factors. Given the availability of two resources, the organism has a negative impact on each resource, and the sum of those two vectors determines the overall impact vector of the organism on the resources (fig. 2.5a). When there are two predators, the organism has positive impacts on both predators (fig. 2.5b). In the presence of a resource and a predator, the organism has a negative impact on the resource and a positive impact on the predator (fig. 2.5c). Finally, when there are a limiting resource and a stress agent, the organism has a negative impact on the resource and no impact on the stress agent (fig. 2.5d). In each case, the slope of the sum impact vector illustrates the proportional relative magnitudes of the organism's impact on the two environmental factors; the size of this vector can be scaled to allow comparisons of absolute impact size when more than one type of consumer is considered. Again, although we have focused on a few modules, impacts can be derived for any combination of factors.

Some Caveats

We have described cases where both the proportional contributions to fitness (requirements) and the relative per capita impacts are fixed. Empirical studies indicate that these relationships are not always linear or additive. In fact, there can be a wide variety of patterns in the relative requirements and impacts of organisms. For example, in the case of resource competition, Tilman (1982) has identified at least six qualitatively distinct classes of resource utilization patterns, many of which are nonlinear (e.g., fig. 23 in Tilman 1982). In chapter 6 we will consider some of these and other causes and consequences of nonlinearity.

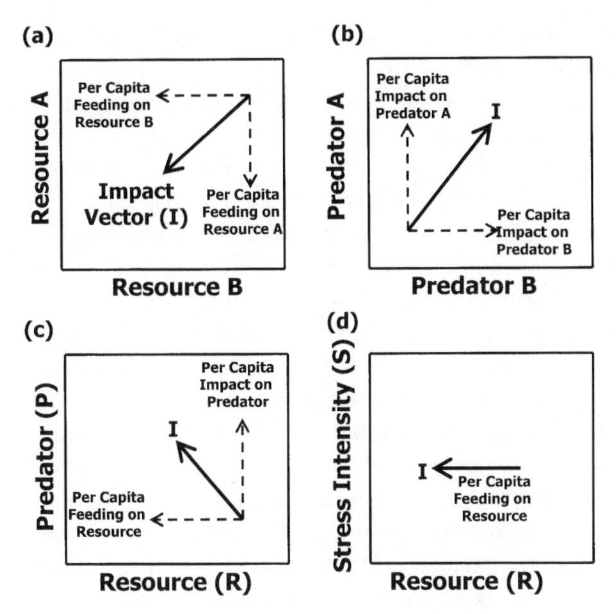

Fig. 2.5. Examples of four modules of the impact niche of a species on two factors. (a) Two resources. (b) Two predators. (c) A resource and a predator. (d) A stress and a resource. Broken lines represent the impact on each factor independently, while the solid line represents the sum of the vectors and is labeled as the impact vector (I).

One might imagine that it would be straightforward to extend the approach described above for two factors onward to address three or more factors. Hutchinson (1957) placed much emphasis on his view of the n-dimensional hypervolume as a metaphor for the niche. Although we do not object to this approach, we will downplay it for several reasons.

First, a tremendous appeal of the approach we advocate is that much (though not all) of the analyses can be done graphically in two dimensions. Except for those fortunate enough to be able to visualize graphs in four or more dimensions, this becomes increasingly difficult. Under these conditions, one might as well take an analytic approach (e.g., Huisman and Weissing 2001a,b).

Second, the approach we advocate is most useful in hypothesis formation and testing. Two factors are the minimum needed to address hypotheses that involve trade-offs (which we discuss more fully in chapter 3), and so the ability to address two factors is important. However, as in previous work on the niche, the idea becomes less useful, and can border on the tautological, when an infinite number of possibilities are included. Under these cases, one can rapidly get caught up in a Sisyphean search for *the* important combination of factors that affect a species, and many of the justifications for taking this simplified approach are weakened.

Finally, as we have suggested, organisms respond to and impact any number of different types of factors. However, many questions of interest to ecologists involve the coexistence of species that utilize factors in similar ways and the ways ecological patterns respond to gradients of an environmental factor. For these sorts of questions, it is often not necessary to consider every factor that influences a species or group of species. For example, Liebig's law of the minimum (Liebig 1840) suggests that the growth rate of a species is determined by the resource that it finds most limiting. While Liebig's law was developed for plants consuming limiting nutrients, this basic principle should hold for any factor that an organism finds limiting. That is, while there are always a multitude of environmental factors that interact simultaneously to influence an organism's growth rate, many of these factors can reasonably be considered as background within a given environment, so that one or two factors are of primary importance in any given situation. We return to this issue in chapter 4 in our discussion of approaches to empirical measurement of these processes.

2.3. A Theoretical Basis for Defining the Niche

In the discussion above we identify a simple but useful way of describing what a species requires from (fig. 2.4) and the impacts it has on (fig. 2.5) its environment. If we envision the niche as the relationship between an organism and its environment, then both of these components are crucial in defining the niche concept. Previous attempts at niche-theoretic definitions had their shortcomings because they focused on one or the other of these two components without a clear synthesis of the two. For example, Hutchinson's definition focused on requirements and did not do much to account for the ways organisms affected their environment. In contrast, much of the niche theory (see, e.g., Vandermeer 1972) developed in the 1960s and 1970s focused on resource utilization functions that described how organisms removed resources from the environment (i.e., the competition coefficient in the Lotka-Volterra equations) without accounting for how such resource use translated into fitness. That is, this body of theory focused primarily on impacts.

Of course, requirement and impact should be closely related. For example, if assimilation efficiency on every resource were identical, then requirement and impact would be related in a one-to-one manner (see appendix for analytical justification). In such cases, the validity of previous work (i.e., work focused on resource utilization functions) will be upheld by our approach. This is because under restrictive circumstances, the formalism of our focus on requirements and impacts converges with the Lotka-Volterra models (see chapter 3), although the latter are based on phenomenological parameters while the former are based on mechanistic parameters. However, when there is not a one-to-one relationship between requirement and impact, then many other possibilities arise. This leads us to propose the following formalization:

The niche is described by the ZNGI of an organism along with its impact vectors in the multivariate space defined by the set of environmental factors that are present.

2.4. Interspecific Interactions and Coexistence in Community Modules

Above, we focused on situations involving direct interactions between an organism and its environment. However, most species in any given natural community interact with numerous other species, and understanding these interspecific interactions is at the heart of ecology. Species sharing niche factors will interact via indirect mechanisms. For example, under resource competition, a species impacts resources, which in turn reduces the amount of resources available to the other species. Under these conditions, pairs of species will generally have reciprocally negative interactions (i.e., show competition in the broad sense of the term). Fig. 2.6 is a qualitative depiction of each of five modules of species interactions that we discuss below. We emphasize that many other modules, such as the incorporation of mutualisms, should provide equally interesting but qualitatively similar conclusions.

In order to consider whether species will coexist or whether one or the other will dominate, we need to consider their ZNGIs, their impacts, and the environment simultaneously. To depict the environment, we draw a *supply vector*, which can lie anywhere in the state space, but is generally examined at the ZNGI. The supply vector points in the direction that the factor being impacted would tend to progress in the absence of species. The point at which the supply would no longer change is called the *supply point*, which can be plotted in two dimensions for any factor of interest. Thus, the supply of consumable resources (fig. 2.7a,b) is the environmentally determined amount of resources as determined by the resources' own limitations. The supply of predators (fig. 2.7b,c) is generally zero in the absence of prey (but it could be nonzero if predators have alternative resources). Finally, the supply of stress (fig. 2.7d) is determined by the abiotic environment; it is not affected by the presence of consumers.

Below, we present each module separately. We will focus the most detail on the resource competition module for no better reason than

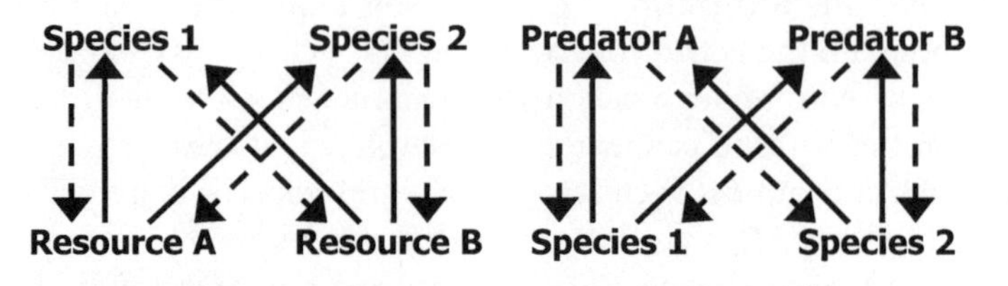

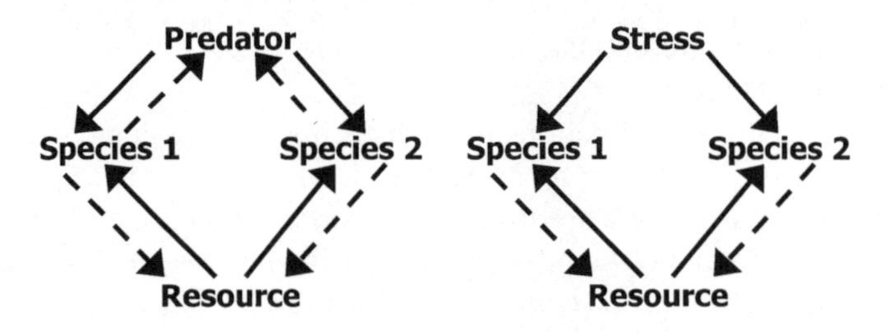

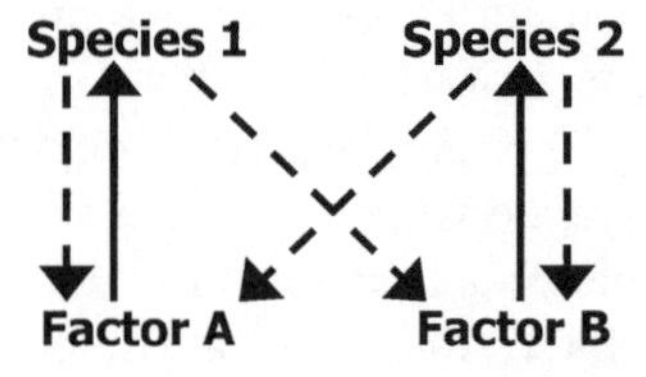

Fig. 2.6. Interaction modules that we will consider in this chapter. Each module has two species, which respond to and impact the factors listed. Response components are indicated with solid arrows pointing to the responding species, while impact components are indicated with dashed arrows points to the factor being impacted.

that it has been the best developed and is thus the most well known. However, it should become clear that the general features of each module are qualitatively quite similar. That is, regardless of the types of factors that are most limiting, and the mechanisms of interspecific interactions, the basic qualitative conclusions from each module are quite robust. As such, this conceptual approach can greatly simplify what might seem to be an insurmountably complex problem. Although here we focus on their graphical depictions, we present the simplest analytical versions of some of these models in the appendix to this chapter.

Resource Competition

The most well known case of competition and coexistence comes from the resource competition module developed by MacArthur

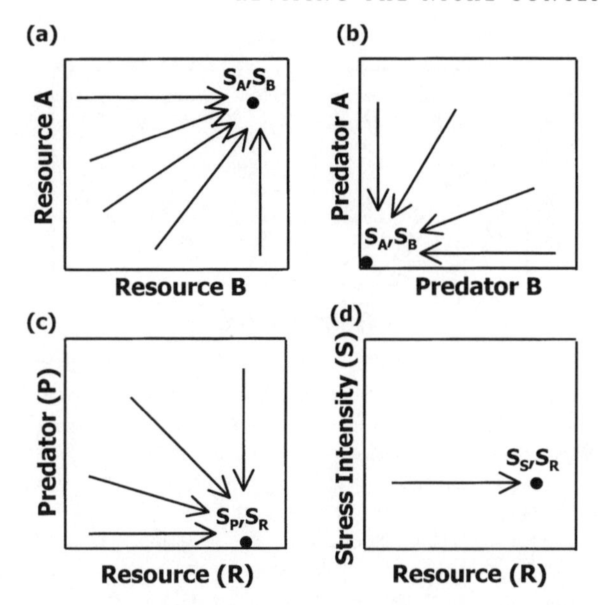

Fig. 2.7. Supply vectors (arrows) and supply point (filled circle) for a given environment of two factors. (a) Supply point and vectors for two resources A and B. (b) Supply point and vectors of two predators A and B (depicted here as being unable to survive in the absence of prey). (c) Supply point and vectors of a predator (P) and a resource (R). (d) Supply point and vectors of a stress (S) and a resource (R).

(1972) (see also Tilman 1982, 1988). When two species compete for two substitutable resources, local coexistence requires three conditions (fig. 2.8).

First, their ZNGIs must intersect. If the ZNGIs do not intersect (not illustrated), then one species will be uniformly better at reducing both resources, will outcompete the other species, and will exist alone, regardless of the impact vectors, so long as the supply point lies above the ZNGI for the better competitor. If the ZNGIs do intersect, this means that one species has a lower resource requirement (lower intercept = R*) for one resource, but has a higher resource requirement (higher intercept = R*) for the other resource. Another way of saying this is that the species with the lowest intercept for the resource along the y-axis has the shallowest sloped ZNGI. Biologically, this means that the species show a trade-off in their ability to exploit the two resources; one species does better on one resource type, but worse on the other, while the second species shows the opposing pattern. Finally, although not explicitly considered here, imagine that the two ZNGIs are identical in the entire state space. This would be analogous to Hubbell's (2001) neutral competitor scenario. In this case, one species or the other would eventually go extinct in the absence of some other factor (e.g., spatio-

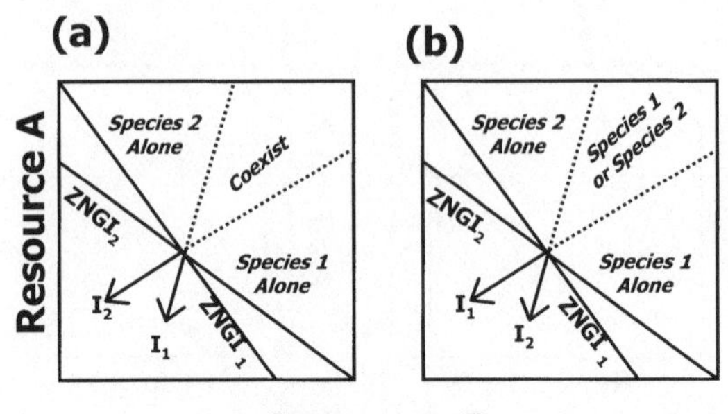

Fig. 2.8. ZNGIs and impact vectors for two species (1 and 2) consuming two substitutable resources (A and B). If the ZNGIs do not intersect, the species with the lower resource requirements exists alone (not shown). Impact vectors for each species are labeled I and are examined only at the ZNGI intersection to examine the stability of the interaction. Dotted lines are the inverse of the impact vectors; supply points interior to these indicate those where both species could potentially exist, whereas supply points above or below this range will result in competitive exclusion of one species by the other. Ranges of supply points for the different outcomes of dominance or coexistence are labeled. (a) Each species has a larger impact on the resource that most limits it, such that the equilibrium is potentially locally stable. (b) Each species has a weaker impact on the resource most limiting to it, and the interior equilibrium is locally unstable.

temporal heterogeneity) that would allow it to maintain itself (see also Chesson and Huntly 1997; Chesson 2000b).

Second, for the intersection of the ZNGIs to be a stable equilibrium, such that the two species may coexist locally, the slopes of the impact vectors must be correlated with the slopes of the ZNGIs. That is, the species with the ZNGI with the shallower slope (lower requirements for the resource on the *y*-axis) has a relatively stronger impact on the resource by which it is most limited (the resource on which it has a relatively higher R*) (fig. 2.8a). Thus, a second trade-off is required for local coexistence, and this is likely to result if the species forage optimally or are adapted in such a way that they expend more effort foraging on the resource type that they find more limiting (Tilman 1988; Gleeson and Tilman 1992; Vincent et al. 1996). Alternatively, there are a variety of biologically plausible scenarios where the species that is less limited by a particular resource has a greater impact on that resource; this is especially possible when traits related to requirements and impacts are correlated (see appendix for specific conditions and their potential correlations) (fig. 2.8b). Here, the species with the shallower-sloped ZNGI has a relatively stronger impact on the resource by which it is least limited (the resource on which it has a relatively lower R*). While in this scenario there is an equilibrium point so long as the ZNGIs intersect, that equilibrium will be an unstable saddle point. For now, we will primarily focus on the former case, where there is a stable equilibrium. We return in more detail to several types of biological complexities that could lead to unstable equilibria in chapter 5.

Third, in addition to the existence and potential stability of the equilibrium between two species, we also need to consider the environmentally determined supply of the two resources. If the supply point is below both species' ZNGIs, of course, neither can exist, whereas if the supply point is below one species' ZNGI but not the other's, than the latter species will exist alone. When the supply of resources is above both species' ZNGIs, we can predict the outcomes of interspecific competition, and whether the species will coexist or one will dominate, by considering the supply in relation to the position of the two species' impact vectors. The two species will coexist locally (provided that the first two criteria are met) only if the slope of the supply vector (and associated position of the supply point) is intermediate in its ratio of the two resources. This is depicted graphically by extending the slope of the impact vectors away from the origin; resource supplies between the two impact vector slopes denote the range of supply where the two species can potentially coexist. In this case, each species is more limited by one resource or the other, they have greater impacts on the alternative resource, and they can coexist stably (center portion of fig. 2.8a). Another way of saying this is that intraspecific competition is greater than interspecific competition—a criterion that converges with the Lotka-Volterra models of interspecific competition (see chapter 3 for a more extended discussion of this comparison). Alternatively, when the supply of the two resources is intermediate but the species' impact vectors are positioned such that the equilibrium is unstable, as in fig. 2.8b, one species or the other will exist alone depending on their initial densities. Another way of saying this is that interspecific competition is greater than intraspecific competition, and the predicted unstable equilibrium again converges with the traditional Lotka-Volterra models.

If, on the other hand, the supply is more strongly biased toward one resource type or the other and is above or below the range of coexistence depicted by the inverse of the impact vectors, then one or the other species will competitively exclude the other (fig. 2.8a,b). Biologically, this means that when a supply vector is strongly skewed toward one resource or the other, the species that is a superior competitor for the resource in greater supply will be able to sequester those resources and build its numbers, allowing it to competitively exclude the species that is an inferior competitor for those resources (for empirical examples of this phenomenon, see Tilman 1982; Chase 1996b; Grover 1997).

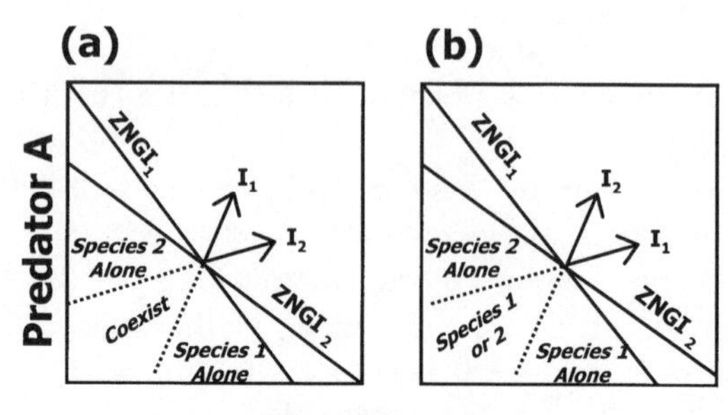

Fig. 2.9. ZNGIs and impact vectors for two prey species (1 and 2) that have two common predators (A and B). if the ZNGIs do not intersect, the species that can survive with the higher density of predators will exist alone (not shown). ZNGIs, impacts, and ranges of supply where one or the other dominates or they potentially coexist are analogous to fig. 2.8. (a) The impact vectors are such that each species has a stronger impact on the predator to which it is more vulnerable, and there is a range of stable coexistence. (b) The impact vectors are such that each species has a greater impact on the predator to which it is less vulnerable, and the interior equilibrium is locally unstable.

Shared Predators

As with all of the modules, the first criterion for two species to coexist when they share two predators is that their ZNGIs must intersect (fig. 2.9) (see Leibold 1998 for more detail). In this case, the intersection of the ZNGIs indicates that the species trade off in their susceptibility to the two predators; one prey species is less affected by one predator (higher P*), and the other prey species is less affected by the second predator. If this does not occur, and one species has its ZNGI above the other (a higher P* for both predators), that species will exist alone by being a superior "apparent competitor" (sensu Holt 1977; Holt et al. 1994). Likewise, each species must have a greater impact on the predator by which it is less affected (fig. 2.9a). Otherwise, the ZNGI intersection will represent an unstable point leading to alternative stable equilibria (fig. 2.9b). Finally, we also need to consider the position of the supply point and vector of predators. Typically, the supply point of two predators will be the origin, since neither predator could exist in the absence of its prey, and the two prey will coexist. However, there could be some external "supply" of predators, such as immigration from nearby habitats, that could cause the density of one predator or the other to be high enough so that one prey species would be driven extinct and the other would exist alone.

Shared Resources and Shared Predators

The simplest form of this module depicts the interactions between two species that share a food resource and are also consumed by a common predator (fig. 2.10) (see Holt et al. 1994; Leibold 1996; Chase, Leibold, and Simms 2000). Here, the first criterion for coexistence (intersecting ZNGIs) is satisfied if they trade off in their relative ability to consume resources and susceptibility to predators. If the ZNGIs do not intersect, the species that is able to survive with the lower level of resources regardless of the abundance of predators will exist alone. That is, the more defended prey species (steeper ZNGI) must incur a cost that makes it a poorer resource competitor (higher R^*). Such a trade-off was first documented by Paine (1966) and has been repeatedly observed in a wide variety of systems (e.g., Dodson 1970; Hall et al. 1970; Janzen 1970; Connell 1975; Simms 1992; Skelly 1995; Kraaijeveld and Godfray 1997; Leibold and Tessier 1997; Bohannan and Lenski 1999, 2000). Second, for the equilibrium to be stable, the species' impacts must be positioned such that the better-defended species (with the steeper ZNGI) has a shallower impact on predators and the better competitor species (shallower ZNGI) has a steeper impact on predators (fig. 2.10b). Biologically, this means that the species that is more affected by predators also provides better food for predators. If, on the other hand, the species that is better defended against predators is also better food for predators, the equilibria will be unstable, and the species will not coexist locally (fig. 2.10b). Finally, as with the other modules, we need to consider the position of the supply. Since predators are not expected to have a supply independent of the prey, we can generally view the supply as the resource supply point. When the equilibrium is stable (fig. 2.10a), the resource supply must be between the impact vectors, where each species is more limited by a different factor, in order for the two species to coexist. Regardless of whether the equilibrium is stable or unstable, when there is low resource supply, the better competitor (the one less limited by resource availability) will be able to outcompete the better-defended prey species and exist alone. Alternatively, at high resource supply, the better-defended species (the one less limited by predators) will be able to outcompete the better resource competitor and exist alone. A variety of alterations and complexities have been incorporated into this basic module (Holt et al. 1994; Leibold 1996; Grover and Holt 1998; Chase 1999a; Chase, Leibold, and Simms 2000), some of which will be discussed in chapter 5.

Shared Resources and Shared Stresses

Species interactions involving a shared resource and a shared stress agent, such as acidification or toxic chemical pollution, differ slightly from the three modules discussed above. This is because we assume

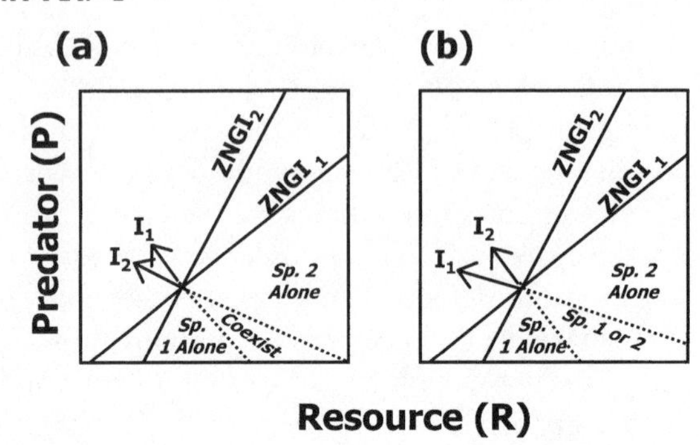

Fig. 2.10. ZNGIs and impact vectors for two species (1 and 2) that share a resource (R) and a predator (P). If the ZNGIs do not intersect, the species that can survive with the lower level of resources and higher level of predators will exist alone (not shown). ZNGIs, impacts, and ranges of supply where one or the other dominates or they potentially coexist are analogous to fig. 2.8. (a) The impact vectors are such that the species most limited by predators has a greater impact on predators, and the species can coexist locally at intermediate resource supply. (b) Impact vectors are such that the species that is least affected by predators has a greater impact on predators, and the interior equilibrium is locally unstable.

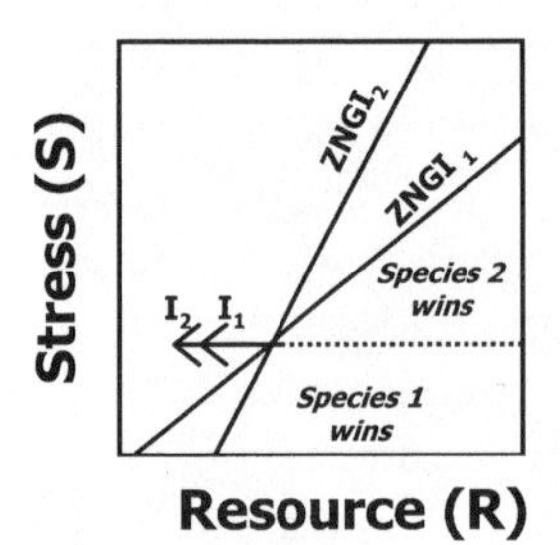

Fig. 2.11. ZNGIs and impact vectors for two species (1 and 2) that share a resource (R) and a stress (S). In this case, so long as the species do not impact stress, they will never coexist locally. Instead, whichever species can maintain resources at the lowest level for a given level of stress will exist alone.

that neither species can have an impact on the stress agent itself and that the impact vectors have identical (flat) slopes, since they alter only the resource density but not the level of stress. If the ZNGIs do not intersect, the species that is able to maintain the lowest level of resources, regardless of the level of stress, will exist alone (not illustrated). Alternatively, the ZNGIs will intersect if one species is a more efficient competitor (lower R^*) without stress, but is more affected by stress (shallower ZNGI) than the other species (fig. 2.11). Since the species have impact vectors with identical slopes, the magnitude of the impact does not influence the competitive outcome at equilibrium. Consequently, since there is no linked trade-off resulting in each species having a greater proportional effect on the factor that it finds most limiting, the species cannot coexist locally regardless of the supply of resources and stresses. Instead, the species that is more tolerant of the stress exists alone when the supply of stress exceeds the amount where the ZNGIs intersect, whereas the species that is less tolerant of stress but a better resource competitor exists alone when the supply of stress is less than that amount.

Bioengineering and Other Nontrophic Interactions

Thus far, we have focused primarily on trophic interactions. However, many organisms can alter the environment in ways that do not (at least directly) affect their own fitness but do affect the fitness of other organisms. For example, large mammals can often alter the fitness of some plant species by their movements (e.g., trampling, wallowing) even though they do not consume those particular species and the density of these plant species does not affect the large mammal's fitness. A consequence, however, is that such effects on vegetation could alter the fitness of other species. Jones et al. (1997) have examined a variety of such interactions, which they have termed "bioengineering," and argued that they are common and important (see also Jones and Lawton 1995). Although such cases seem to be quite distinct from the trophic interactions we have discussed so far, they can also be examined in this niche-based framework.

Imagine a case where two species (species 1 and 2) each consume a distinct resource (resources A and B respectively) and that either one or both of these species has a negative incidental impact on the resource it does not consume (for example, a forager trampling the resource of another forager) (fig. 2.12). Because they consume distinct resources their ZNGIs are orthogonal to each other (i.e., at ninety-degree angles to each other). In the cases of fig. 2.12a,b, one species (species 2 in 2.12a, species 1 in 2.12b) does not impact the resource of the other species and has an impact vector orthogonal to its own ZNGI. However, the other species does impact the resource of the first species, and its impact vector has a positive slope. In the case of fig. 2.12c, each species impacts the other's resource. The resulting interactions can be analyzed in just the same way as we did above. We conclude that two criteria for coexistence, intersecting ZNGIs and correlations between the slopes of the ZNGI and the impact vectors, are easily met. This result is perhaps not surprising, since the species do not directly share resources. However, coexistence is not guaranteed, since one or both species can negatively impact the other's resource through bioengineering. Coexistence in this situation is thus most dependent on the relative renewal rates of the resources so that the supply vector is intermediate in slope to the two impact vectors.

Alternatively, several cases of bioengineering will not negatively impact the resource of another individual but can influence the resource

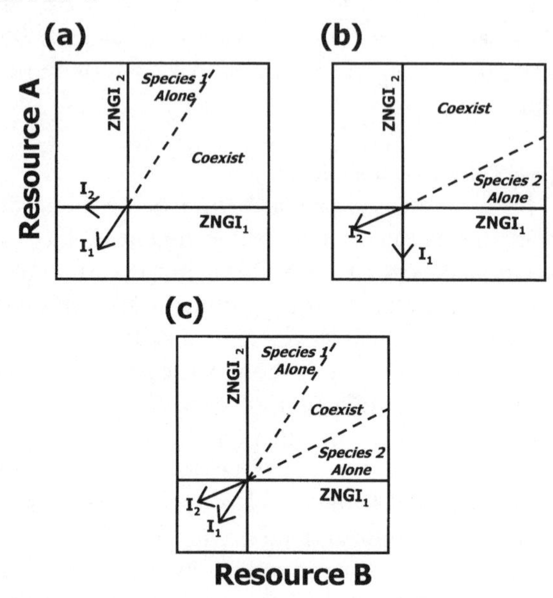

Fig. 2.12. ZNGIs and impact vectors (I) for two species that do not share a resource but can impact each other's resource (i.e., bioengineers). (a) Species 1 impacts species 2's resource, but not the reverse. (b) Species 2 impacts species 1's resource, but not the reverse. (c) Each species impacts the other's resource. Ranges of resource supply where each species exists alone are labeled. Coexistence is likely, but a species can exclude the other with sufficient quantities of its own resource.

supply (Jones et al. 1997). For example, beavers are bioengineers that create habitats for a number of species that rely on ponds created by beaver dams. Additionally, waste products from feeding fish can provide nutrients that influence algal growth (e.g., Vanni, Layne, and Arnott 1997; Vanni, Flecker, Hood, and Headworth 2002). These sorts of bioengineering affect supply, which can be easily analyzed by examining how the level of supply of a particular resource varies as the density of the bioengineer varies.

2.5. Local versus Regional Criteria for Coexistence

Interactions at the local scale occur via effects of species on each other's per capita birth and death rates. The conclusions derived above are that coexistence at a local equilibrium involves the three criteria of (1) intersecting ZNGIs, (2) distinct impact vectors that result in each species having a greater relative impact on the environmental factor that most limits its own population growth, and (3) a supply vector (and associated supply point) that is between the impact vectors of the competitors. However, species can interact with each other via effects that occur at larger spatiotemporal scales.

A region consists of several localities, each of which has conditions that favor either existence or exclusion of one or both species. Species that cannot coexist locally could conceivably coexist regionally by existing in distinct local communities within the same region, and thus the criteria for regional coexistence are simpler than those at the local level (Leibold 1998). For both local and regional coexistence, criterion 1 above must always be met, as each species must be favored in at least one locality. However, criteria 2 and 3 are not necessary for regional coexistence. If criterion 2 is not met, then the local equilibrium will be unstable, alternative stable equilibria will occur, and one or the other species will be present alone in a local community. In this case, because there are many local communities in a defined region, each of which contains one consumer or the other existing alone, the two species may be able to coexist regionally. Finally, criterion 3 is not necessary for regional coexistence so long as there is heterogeneity in the local environmental conditions (supply), such that each species has at least one locality where it is predicted that it will exist. We will expand on these ideas of local and regional coexistence criteria to multispecies communities in chapters 7 and 8.

2.6. Trade-offs

In all of the modules discussed above, we emphasized that two species must differ in both their requirements and impacts in order to potentially coexist locally and in their requirements in order to potentially coexist regionally. That is, the species must show trade-offs in order to coexist in these simple modules. Trade-offs occur when a genetic, physiological, morphological, behavioral, or ecological trait of an organism that confers advantage for performing one function simultaneously confers disadvantage for performing another function. The search for trade-offs in nature has been somewhat contentious, particularly at the within-species level (e.g., Stearns 1992; Bergelson and Purrington 1996; Whitlock 1996). There are both theoretical and empirical explanations why trade-offs would be hard to detect, or absent, in fine-scale within-species studies. However, trade-offs among species are much easier to detect, especially when the species involved come from different positions along important environmental gradients (i.e., do not necessarily coexist in local communities).

Even in among-species studies, there is some controversy about the prevalence of trade-offs. Both classical theory and our exploration of mechanistic niche-based models show that species must show some degree of differentiation and trade-offs in order to coexist; this is known as Gause's axiom (see Hardin 1960; Levin 1970). However, some have viewed Gause's axiom as tautological, because of a potentially never-ending search for the niche axis along which two species might differ (Hubbell 2001). When trade-offs are not found along the niche axis considered in a study, Gause's axiom cannot be rejected, because trade-offs may occur along other, unmeasured axes. Such an approach is indeed problematic. As we discussed in chapter 1, Hubbell (2001) has suggested that a community can maintain high species diversity for long periods of time, while many species are actually on a slow trajectory toward extinction (also see Bell 2000, 2001), without adhering to Gause's axiom. That is, Hubbell's hypothetical species show no niche differences or trade-offs.

Gause's axiom could be less tautological if one could test the null hypothesis, that there are no trade-offs, without identifying the specific niche axes on which species trade off. In fact, this is a relatively straightforward task to accomplish by experimentally reducing the abundances of each of the interacting species and following their resultant population trajectories (a "pulse experiment"; Bender et al. 1984). If a community returns toward a similar configuration after a system is perturbed, then it can be concluded that there is some sort of attractor (whether it be a stable point, an oscillatory trajectory, or a chaotic attraction) that allows the species to coexist in the long term. We argue that such an attractor in a closed system is prima facie evidence that a trade-off exists even though the exact mechanistic basis of this trade-off is not necessarily specified. This test has been performed a number of times to explore the interactions and coexistence of species (see reviews by Connell 1983; Schoener 1983; Sih et al. 1985; Gurevitch et al. 1992; Resetarits and Bernardo 1998; Gurevitch et al. 2000). That is, if all species increase when reduced to lower densities, then they likely exhibit some sort of trade-off that allows them to coexist in a multispecies community. Following such a rejection of the null hypothesis that there is no trade-off, we would then argue that a search for the mechanism of niche differentiation would be justified even if it were to take numerous attempts.

2.7. Concluding Remarks

The explicit recognition of both requirement and impact components of the niche provides a general framework for understanding the problems of species interaction and coexistence, which form the basis of much ecological inquiry (see also Leibold 1995). In addition, these sorts of models have heretofore primarily focused on interspecific competition for limiting resources, whereas we show that the same basic principles in those models apply to other modules of species interactions. In the remainder of this book, we build upon this basic framework to address a variety of empirical and theoretical problems. While both empiricists and theoreticians might want more detail or precision, it is our view that these simple graphical models and their underlying analytical structure provide a particularly lucid way to make broad comparisons and general predictions.

With similar goals of achieving a synthetic general framework for understanding species interactions and coexistence, Chesson (2000b) took a distinct but conceptually quite similar approach. Specifically, Chesson showed that the factors promoting coexistence can be classified as "equalizing," which makes species similar to each other in some way, and as "stabilizing," which affects the feedback in the system so that perturbations result in a return to a stable equilibrium. In our framework, coexistence requires intersecting ZNGIs, which is in Chesson's terminology "equalizing," in that it ensures that both species have some advantage (i.e., neither species is a "dud" in relation to both factors). Coexistence also requires a correlation between ZNGIs and impact vectors, which is stabilizing in that it creates a negative feedback between the species impacts and requirements. As a result, neither species can decrease when rare or increase when common. Finally, coexistence requires that the supply of the factors be intermediate to the species impacts, which is equalizing in the sense that the local environment does not disproportionately favor one species over the other. Both equalizing and stabilizing effects are required for coexistence even in situations that don't go to a point equilibrium (Chesson 2000b), which we return to in chapter 6.

The definition of niche has traversed a turbulent road since its inception by Grinnell and Elton and its deification by Hutchinson and MacArthur. Confusions about the niche concept have sparked a number of reviews (Hutchinson 1978; Schoener 1989; Griesemer 1992), calls for redefinition (Whittaker et al. 1973; Maguire 1973; Hurlbert 1981; Colwell 1992), and calls for dismissal of the concept altogether (Williamson 1972; Margalef 1974; Bell 1982; Hubbell 2001). The niche concept has also piqued the creativity of some of the more poetic ecologists that illustrate these frustrations (all appear in Hurlbert 1981):

> Let's consider the concept of niche—
> If I knew what it meant I'd be rich.
> Its dimensions are n
> But a knowledge of Zen
> Is required to fathom the bitch.
> —Grant Cottam and David Parkhust

With your concept of niche I agree
But there's clearly one hitch I can see
You blame the wrong sex
For the inherent hex,
For the niche is no she, but a he.
—Joy Zedler

I'm amazed that a smart woman like Joy
Would believe that the niche is a boy;
For a niche is elusive,
Deceitful, confusive—
It's quite clear it's a feminine ploy.
—Grant Cottam

We believe that the niche concept plays a pivotal, albeit frustrating, role in a variety of important ecological questions. Our goal in this book is to provide a link between mechanistic models and classical concepts. We would thus answer Cottam and Zedler with our own poem (which also explains why we are scientists and not poets):

The contributions of the niche are not stingy
but its definitions have been rather dingy.
The framework that's needed,
With gender issues unheeded,
requires not Zen, but a ZNGI.

2.8. Summary

1) We propose a quantitative definition of niche that includes a species' requirements from and impacts on a factor. Requirements refer to the conditions where an organism's birth rates exceed its death rates on a particular factor, whereas impact refer to the effect that species has on the factor.
2) The niche concept can be extended in two dimensions to show how factors interact to determine the two components of the niche of an organism. This framework is flexible and can be used for any factor that influences the fitness of an organism or on which that organism has an impact.
3) Species that utilize resources in similar ways can be examined on similar axes to discuss the conditions necessary for coexistence.
4) Regardless of the basic limiting factors (i.e., resources, predators, stresses) species use, in order for species to coexist locally, they

must show linked trade-offs. First, their zero net growth isoclines (ZNGIs) must intersect such that they require resources differentially. Second, their impact vectors must be positioned so that each species has a relatively greater impact on the resource that it finds most limiting. If the reverse is true, the equilibrium will be unstable. Finally, the supply of the factor must fall intermediate to the impact vectors. If it does not, it will disproportionately favor the species that is best able to utilize the most abundant factor.

5) The criteria for regional coexistence are easier to achieve and require only that species show trade-offs in their requirements and that environmental supply rates be sufficiently heterogeneous across space.

Appendix

In this chapter, and throughout the book, we emphasize graphical models. However, we fully recognize the limitations of completely relying on the graphical models and emphasize that a more complete theoretical treatment of much that we discuss requires mathematical foundation. Here we give a brief overview of the mathematical foundation that underlies the simplest community modules discussed in this chapter. The mathematics for these modules have been discussed frequently previously (most notably by MacArthur [1972]; Leon and Tumpson [1975]; Tilman [1980, 1982]; Holt et al. [1994]; Leibold [1996, 1998]; Grover [1997]; Wootton [1998]; Chase, Leibold, and Simms [2000]), but we include them here for reference. We do not incorporate some of the more complex mathematics associated with the issues raised in the following chapters, but instead refer the reader to the primary literature (e.g., Armstrong and McGehee 1980; Gleeson and Tilman 1992; Vincent et al. 1996; Wootton 1998; Abrams 1999; Chase 1999a; Huisman and Weissing 1999, 2001a,b; Chesson 2000b). In addition, we hope that those who are more mathematically inclined will continue to explore expansions and limitations of these simple but powerful models.

We present three modules of species interactions here: resource competition for two substitutable resources, apparent competition for two predators, and keystone predation, where species compete for resources and are consumed by a common predator. There are other modules that have been developed analytically, such as disturbance (Wootton 1998), as well as many others that have not yet been fully developed, such as mutualisms and bioengineering. In all of the formulations discussed, we use the notations and terminology of Leibold (1998), but note that while the notations and formulations might be slightly different from those found elsewhere (e.g., Tilman 1982; Holt et al. 1994; Grover 1997; Chase, Leibold, and Simms 2000), the basic parameters are nearly identical.

Competition for Shared Resources

This is the most often studied model of reciprocally negative interactions among species; here, we discuss competition for substitutable resources, but the same general results apply when resources are essential. The main results for the simplest versions of the model were developed by MacArthur (1972) (see also Leon and Tumpson 1975; Tilman 1980, 1982), and a variety of complicating factors involving nonlinear effects, spatiotemporal variation, population cycling, etc. have been well studied (e.g., Tilman 1982; Abrams 1988; Chesson 1990; Huisman and Weissing 1999, 2001a,b). For generality and simplicity, the linear version of the model is sufficient to illustrate the main points and the fundamental dynamic equations describing how a guild of consumers (whose densities are denoted N_i) compete for two shared resources (whose densities are denoted R_1 and R_2). These models assume that the dynamics of these species can be depicted by a series of simultaneous differential equations, which in this case can be written as

$$dN_i/dt = N_i(f_{i1}a_{i1}R_1 + f_{i2}a_{i2}R_2 - d_i)$$
$$dR_1/dt = c_1[S - R_1] - \Sigma[f_{i1}N_iR_1]$$
$$dR_2/dt = c_2[S - R_2] - \Sigma[f_{i2}N_iR_2],$$

where f_{ij} is the per capita feeding rate of consumer i on resource j, a_{ij} is the conversion efficiency by consumer i of eaten resources of type j, d_i is the per capita loss rate of consumer i (including metabolic respiration and death), and $c_j[S - R_j]$ is the turnover rate of resource j in the supply pool (S) that are not tied up in R.

Given this set of equations, we can derive the resource requirements, as determined by the zero net growth isocline ($ZNGI_{(i)}$) for each species (i). The ZNGI is determined by setting all of the differentials (dX/dt) equal to zero, which assumes that the population is not increasing or decreasing (i.e., population growth is zero) and the species are at equilibrium. The impact vector (I(i)) is determined as the per capita rate of resource consumption for each resource. Specifically:

$$ZNGI_{(i)}: R_2^* = (d_i - f_{i1}a_{i1}R_1^*)/(f_{i2}a_{i2})$$
$$I_{(i)}: [f_{i1}R_1, f_{i2}R_2].$$

The slope of the ZNGI, its intercept on the R_1 axis (this is equivalent to Tilman's [1982] R*), and the slope of the impact vectors can also be derived; they are:

slope of $ZNGI_{(i)} = -f_{i1}a_{i1}/(f_{i2}a_{i2})$
$ZNGI_{(i)}$ intercept on R_1 axis $= d_i/(f_{i1}a_{i1})$
slope of the impact vector $= f_{i2}R_2/f_{i1}R_1$.

Thus, as discussed in chapter 3, if two species, A and B, compete for the two resources (R_1 and R_2), a necessary condition for local coexistence at an equilibrium point between these species is that

$$(f_{B2}a_{B2})/(f_{B1}a_{B1}) > (f_{A1}a_{A1})/(f_{A2}a_{A2}).$$

Verbally, the species with the lowest resource requirements on resource 1 must be relatively less efficient at feeding and converting resource 2 (i.e., there must be a trade-off). Graphically, this means that the species with the lowest requirements for R_1 must have the steepest ZNGI. Additionally, for an equilibrium to be feasible, involving positive densities for all species, there must be constraints on d_i such that the intersection of the ZNGIs occurs where both R_1 and R_2 are positive. Furthermore, if this equilibrium is locally stable, two species, A and B, must differ in the slope of their impact vectors such that:

$$f_{B2}R_2/f_{B1}R_1 > f_{A2}R_2/f_{A1}R_1.$$

Verbally, the species with the shallowest slope to its ZNGI must have the steepest impact vector; each species has a greater relative impact on the resource from which it benefits. Finally, note that the environmentally determined resource supply (S) does not determine whether the equilibrium is stable, but does determine whether or not the stable equilibrium is achieved. If S is below the ZNGIs of the two species, neither will exist, whereas if it disproportionately favors one species or the other, then that one will outcompete the other and exist alone.

We have focused here on competition for two substitutable resources. Although the algebra gets more complex, similar results can be obtained with competition for two essential resources (e.g., Vincent et al. 1996). Similarly, these equations can be extended to more than two resources, although the graphical techniques and algebraic constraints become much more complex (Huisman and Weissing 2001b).

Apparent Competition via Shared Predators

Another possibility is that density-dependent interactions are entirely mediated via shared predators: a mechanism Holt (1977) termed "apparent competition." For simplicity and by analogy with the resource competition model, we assume that each species grows exponentially in the absence of the predators and that the interactions are linear. We further assume that the system is open with regard to predator immigration and emigration in order to add density dependence and stabilize the system (R. Holt, personal communication). Under these conditions the dynamic equations for a guild of such species (whose densities are denoted N_i) interacting with two predators (whose densities are denoted P_1 and P_2) can be modeled as:

$$dN_i/dt = N_i(r_i - f_{i1}P_1 - f_{i2}P_2)$$
$$dP_1/dt = P_1(\Sigma[f_{i1}c_{i1}N_i] - d_1) + (P_1[i_1 - e_1])$$
$$dP_2/dt = P_2(\Sigma[f_{i2}c_{i2}N_i] - d_2) + (P_2[i_2 - e_2]),$$

where f_{ij} is the per capita feeding rate of predator j on species i, c_{ij} is the conversion efficiency by predator i of eaten prey of type j, d_j is the per capita loss rate of predator j (including respiration and death), r_i is the Malthusian parameter for exponential growth in species i and i_i and e_i are predator immigration and emigration rates, respectively. The tolerance of each species to predators and their per capita impacts can be described by a $ZNGI_{(i)}$, and an impact vector ($I_{(i)}$) described by:

$$ZNGI_{(i)}: P_2^* = (r_i - f_{i1}P_1^*)/m_{i2}$$
$$I_{(i)}: [f_{i1}c_{i1}P_1, f_{i2}c_{i2}P_2].$$

Here, the slope of the ZNGI, its intercept on the P_1 axis, and the slope of the impact vectors are:

$$\text{slope of ZNGI}(i) = -f_{i1}/f_{i2}$$
$$\text{ZNGI}(i) \text{ intercept on } P_1 \text{ axis} = r_i/f_{i1}$$
$$\text{slope of the impact vector} = (f_{i2}c_{i2}P_2)/(f_{i1}c_{i1}P_1).$$

The necessary condition for local coexistence of two species (A and B) is that

$$f_{A1}/f_{A2} < f_{B1}/f_{B2}.$$

That is, there must be a trade-off between the two apparent competitors in their susceptibility to predation by each of the two predators. Graphically, this means that the species with the highest tolerance for P_1 must have the shallowest ZNGI. Additional requirements for an equilibrium involving positive densities for all species involve constraints on r_A and r_B such that the intersection of the ZNGIs occurs where both P_1 and P_2 are positive.

Furthermore, if this equilibrium is to be locally stable two species, A and B, must differ in the slope of their impact vectors such that

$$f_{A2}c_{A2}P_2/(f_{A1}c_{A1}P_1) < f_{B2}c_{B2}P_2/(f_{B1}c_{B1}P_1).$$

Thus, the species with the shallowest slope to its ZNGI must have the steepest impact vector; each species has the greatest impact on the predator that most affects its own death rate.

Competition for Shared Resources and Shared Predators: Keystone Predation

Models of reciprocally negative interactions between species that share predators and resources have only recently been developed in a mechanistic way. Though early work (Cramer and May 1971; Vance 1978; Leibold 1989) modeled resource competition using the nonmechanistic Lotka-Volterra formulation, more recent models (Holt et al. 1994; Grover 1994, 1995; Leibold 1996; Chase, Leibold, and Simms 2000) are better suited for discussions related to the niche concept in this book. Again making simplifying linearity assumptions, we can write a model for species (N_i) that compete for shared resources (R) and share a common predator (P) as:

$$dP/dt = P(\Sigma(m_i c_i N_i) - d_P)$$
$$dN_i/dt = N_i(f_i a_i R - m_i P - d_i)$$
$$dR/dt = c[S - R] - \Sigma(f_i N_i),$$

where the notation is the same as in the previous models (subscripts 1 and 2 are deleted since there is only one resource and one predator, and the subscript P refers to parameters for the predator), except that here we add a parameter m_i to indicate the consumption rate of species i by the predator. Here, the resource and predator densities that satisfy each species' requirements and the species' per capita impacts can be described by:

$$\text{ZNGI}_{(i)}: P^* = (f_i a_i R^* - d_i)/m_i$$
$$\text{I}_{(i)}: [f_i R, m_i c_i P].$$

The slope of the ZNGI, its intercept on the R axis, and the slope of the impact vectors are:

$$\text{slope of ZNGI}_{(i)} = f_i a_i/m_i$$
$$\text{ZNGI}_{(i)} \text{ intercept on } R \text{ axis} = m_i/(f_i a_i)$$
$$\text{Slope of the impact vector} = m_i c_i P/(f_i R).$$

In this case, the necessary condition for local coexistence at an equilibrium point between two species, A and B, is that

$$f_A a_A/m_A < f_A a_B/m_B.$$

That is, the species with the lowest resource requirements on resource 1 will be more vulnerable to the predator. Graphically, this means that the species with the lowest requirements for R must have the shallowest ZNGI. Additional requirements for an equilibrium involving positive densities for all species involve constraints on the d_i such that the intersection of the ZNGIs occurs where both R and P are positive. Furthermore, if this equilibrium is to be locally stable the species must differ in the slope of their impact vectors such that

$$m_A c_A P/(f_A R) > m_B c_B P/(f_B R).$$

Again, this means that the species with the shallowest slope to its ZNGI must have the steepest impact vector; each species has a greater impact on the factor that is most strongly limiting. Finally, resource supply (S) must exceed the ZNGIs of the two species and be intermediate for the two species to coexist; otherwise, one or the other will exist alone.

CHAPTER THREE
COMPARING CLASSICAL AND CONTEMPORARY NICHE THEORY

Although the niche framework that we overviewed in chapter 2 derives from a number of contemporary theoretical and empirical studies, it differs substantially from the classical view of ecological niches. The Lotka-Volterra models, upon which much classical niche theory was based, depict competitive interactions among species as a proportional reduction in the density of a focal species by a competitor species (competition coefficient) from the maximum density that the focal species could achieve in the absence of competitors (carrying capacity). Neither the carrying capacity nor the competition coefficient in these models has much direct relation to the underlying biology that creates them, and thus we refer to these models as phenomenological. Even if these quantities were measured, the addition of more species, or variation in the environmental conditions, necessitates a complete remeasurement of the components.

The mechanistic consumer-resource framework on which we focus alleviates some of these problems with the classical niche framework. For example, the outcome of competition between consumers can be predicted with a variety of resource conditions or species with different traits. Furthermore, while many previous treatments have focused almost exclusively on resource competition (e.g., Tilman 1982, 1988; Grover 1997), we provide a synthetic link between resource competition and other types of limiting factors, such as predators and stresses. Finally, by making explicit the two aspects of the niche—requirements and impacts—we clarify a variety of previously confusing conceptual and semantic issues. Nevertheless, some very important and influential work derived from the more traditional niche theory. In this chapter we discuss the similarities and differences between the classical and contemporary approaches to the niche.

3.1. Comparisons between Consumer-Resource and Lotka-Volterra Models

The mechanistic consumer-resource framework overviewed in chapter 2 (our preferred approach) has several commonalities with the Lotka-Volterra competition equations when certain simplifying assumptions are made. MacArthur (1972) and Tilman (1982) have derived some of these commonalities analytically. In chapter 2 we showed that stable coexistence of two species with similar niches requires two linked trade-offs. First, in the absence of the other species, each species must differ in the factor it finds most limiting. Second, each species must have a greater impact on the factor it finds most limiting. If the second trade-off is not upheld, and each species has a relatively stronger impact on the factor that it finds less limiting, then the equilibrium point will be unstable. In this case, only one or the other species will exist locally in alternative stable equilibria. This is equivalent to saying that for two species to coexist locally, interspecific competition must be less than intraspecific competition. Furthermore, this result is consistent with the predictions of Lotka-Volterra competition models:

$$dN_i/dt = N_i r_i (K_i - \alpha_{ii} N_i - \alpha_{ij} N_j)/K_i,$$

where N_i is the species under consideration, r is the intrinsic rate of increase, K is the carrying capacity, and α_{ii} is the intraspecific effect of a species on itself, while α_{ij} is the interspecific effect of species j on species i (with α_{ii} usually set equal to 1). For each of the two competing species the condition for local stability of an equilibrium in this model is

$$\alpha_{ii}\alpha_{jj} > \alpha_{ij}\alpha_{ji}.$$

That is, the (geometric) mean intraspecific per capita effect is greater than the mean interspecific effect. As with the niche-based models described in chapter 2, if this condition is not upheld, an equilibrium, if it exists, will be locally unstable and one or the other species will exist alone depending on initial conditions.

The final condition for coexistence in the niche-based model is that the resource supply not disproportionately favor one species or the other. Similarly, in the Lotka-Volterra model the condition that $\alpha_{ii}\alpha_{jj} > \alpha_{ij}\alpha_{ji}$ is not enough to guarantee coexistence. Vandermeer (1975) showed that coexistence requires that the relative carrying capacities of the two species must be such that neither species has too strong of an advantage. This is expressed in relation to the competition coefficients as

$$\alpha_{ji} < K_j/K_i < 1/\alpha_{ij}.$$

3.2. Relationship to Conventional Niche Theory

Although the approach to niche relations we describe and advocate contrasts in many respects with "conventional niche theory" (CNT) (MacArthur 1969; Vandermeer 1972; Giller 1984). Our discussion above shows that they converge under some conditions, and so much of CNT has relevance to our discussion. However, CNT is fraught with ambiguity and a variety of misconceptions. Our framework allows a more comprehensive understanding of the underlying mechanisms of species interactions and can clear up a variety of misconceptions. Furthermore, our approach is not limited to interspecific competition, and it allows a flexible way to deal with many different types of species traits and interspecific interactions. In the following sections, we use the niche framework developed here to illuminate a series of issues from CNT.

The Fundamental and Realized Niches

One of Hutchinson's biggest insights about niche relations came from making a distinction between the *fundamental* and the *realized* niche of an organism (see chapter 1). The fundamental niche represents all of the conditions in which a species could *potentially* exist, whereas the realized niche represents those conditions in which it *does* exist in the presence of interacting species. The basic insight was that species could exist under a greater range of environmental conditions if their competitors were absent than if they were present (a notion similar to the *niche compression hypothesis*, sensu MacArthur and Levins 1967; Schoener 1974a). The mechanistic framework developed here can help to make these ideas more specific. For brevity, we will illustrate this with the resource competition module

In our niche-based framework, if the resource supply point of the two resources is above the ZNGI of a species, then those resource conditions are part of that species' fundamental niche. That is, the species can potentially exist where the resource densities in the absence of consumption exceed the minimal requirements the consumer species need to maintain positive (or neutral) population growth. Fig. 3.1 illustrates how two consumers overlap in their fundamental niches in all environments whose supply points lie above *both* of their ZNGIs (zones

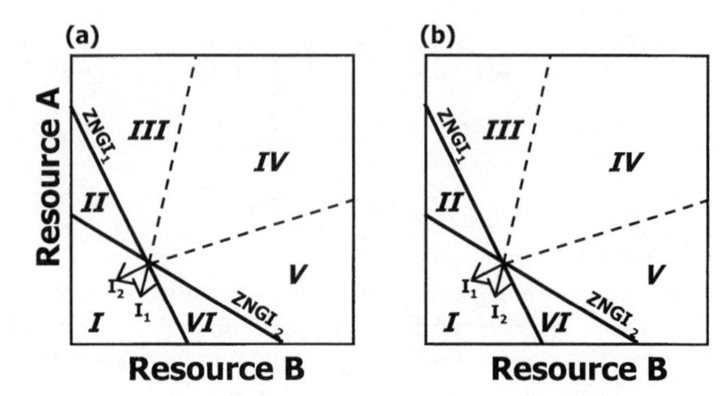

Fig. 3.1. ZNGIs and impacts of two species competing for two substitutable resources. (a) The ZNGI intersection is stable, where each species has a greater impact on the resource that most limits it. (b) The ZNGI intersection is unstable, and each species has a greater impact on the resource that least limits it. Roman numerals indicate zones of resource supply where different qualitative patterns are expected as described in text.

III–V in fig. 3.1). As Hutchinson pointed out, however, the range of resource supply conditions where the two species will actually coexist—their realized niches—is much narrower because competitive exclusion will occur under many resource conditions within each species' fundamental niche.

In fig. 3.1a, we illustrate the situation where the slopes of the impact vectors are positioned such that they produce a stable local equilibrium. In this case, species 1's fundamental niche occurs where the supply point falls in zones III–VI, species 2's fundamental niche occurs in zones II–V, and they consequently have overlapping fundamental niches in zones III–V. However, species 1's realized niche falls in zones IV–VI, and species 2's realized niche falls in zones II–IV; their realized niches overlap only in zone IV, where the graphical model predicts that they can coexist. Consistent with Hutchinson's view, then, coexistence occurs only where the species have overlapping realized niches.

In fig. 3.1b, we illustrate the situation where the slope of the impact vectors produces an unstable local equilibrium, leading to alternative stable equilibria. In this case, the fundamental niches of the two consumers are the same as in fig. 3.1a, but their realized niches are different. Species 1's realized niche is limited to zones V–VI and species 2's realized niche is limited to zones II–III. In zone IV, each species can exist, depending on its initial conditions. When a species is initially rare in this zone, it will go extinct, and the initially more common species will exist alone. Thus, zone IV would not be in either species' realized niche if we define it on the basis of the consumer's ability to invade in the presence of a competitor (i.e., invade when rare).

Alternatively, if we define the realized niche based on the ability of a species to resist exclusion in the face of competition (i.e., prevent the other from invading), each consumer can do so in zone IV so long as its abundance is initially higher than the other species'. In this case, we conclude that both species share zone IV as part of their realized niche but cannot coexist in this shared part. This dependence on initial conditions contrasts with Hutchinson's conceptualization of the realized niche.

Niche Axes

We define a niche axis as a quantitative measure of an environmental factor (e.g., density, concentration, probability of occurrence per unit area). This restriction is essential in our framework because it allows us to express the range of requirements (i.e., locate the ZNGIs) and quantitatively examine the impacts an organism has on the factor relative to its supply. In CNT, however, niche axes have not been as constrained and were often used in fundamentally different ways to index resources (or other factors). For example, niche relations involving granivorous birds have often used seed size as a niche axis (e.g., Cody 1973). Although this is a quantitative metric, it does not measure the occurrence of any single factor and appears inconsistent with Hutchinson's definition since seed size alone is not a metric that can determine if resource requirements are met (in what way would birds compete for seed size?). Instead, such niche axes serve as a means of collapsing a large number of quantifiable resource axes (e.g., density of seeds in size classes x–y) into a single axis. This mathematical *convenience* was at the heart of MacArthur and Levins's (1964, 1967) approach to CNT. However, the conception that this use of niche axes is consistent with Hutchinson's definition is, we believe, a major source of confusion. While there are numerous valid insights that have been drawn from CNT using this approach, it is also fraught with ambiguity and difficulties. Abrams (e.g., 1975, 1988) in particular has emphasized that much more complex dynamics can emerge than one might hypothesize from these models. Alternatively, they can serve as useful guides to help conceptualize the results of more complete analyses (e.g., Brown and Vincent 1987, 1992; Brown 1996).

Niche Overlap, and How It Relates to the Intensity of Competition

Ecologists interested in species coexistence during the 1960s and 1970s spent a lot of time measuring the overlap in the resource utilization of similar species. This resulted in debates about the correct way to measure the diet overlaps of species (e.g., Colwell and Futuyma 1971; Pianka 1973). From this, niche overlap became synonymous with competition intensity. Petraitis (1989) showed that the index of niche overlap derived by Pianka (1973) was equivalent to the cosine of the angle between the impact vectors in resource competition modules (see chapter 2). Using our definition of the niche, niche overlap, as measured from diet overlap studies, represents only one component of competition among species: short-term per capita impacts. Such studies say nothing about the long-term effects of organisms on the abundance of their resources (described by the ZNGIs) and very little about the overall importance of interspecific competition.

The Niche as a Property of the Species and the Environment

This philosophically oriented distinction long ago alerted ecologists to some of the problems associated with definitions of the niche concept. Hutchinson's formal definition of the niche seemed to indicate that he thought the niche was a property of the species (i.e., no matter what the environment was, the ability of a species or population to grow could be determined). In our approach we also focus extensively on aspects of the niche that are strictly a property of the species (i.e., requirements [ZNGIs] and impacts). However, whether a species can exist in a given place also depends on the environment. Thus, our framework considers the fundamental niche as a property of the species (the position of the ZNGIs) as well as the environment (the position of environmental supply). In addition, because the realized niche depends on the species' requirements and impacts, the environmentally determined supply point, and interactions with other species, it too is a property of the interaction between organisms and environment.

It is also important to note that under our definition, both the requirement and the impact component are determined both by the traits of the species in question and by their surrounding environment. For example, the ability of a species to maintain zero net growth on a particular resource will often depend on its physiological responses to the surrounding environmental conditions (e.g., temperature, pH). Similarly, the impacts of a species on a resource may vary from site to site as environmental conditions change. Thus, both ZNGIs and impacts are properties of the interaction between species and their environment.

Empty Niches

The concept of "empty" niches is largely incompatible with the view that the niche is the property of a species (the property cannot exist without the entity that defines it). However, ecologists have often wondered if niche relations could be used to explain the diversity, the saturation, and the invasibility of communities. To do so, they have often invoked (perhaps mostly in metaphorical terms) the empty niche concept. Our framework allows us to recast this question in ways that hark back to differences in the niche concepts of Grinnell and Elton (antedating Hutchinson's definition and usage).

Empty niches can be envisioned as two different, though related, things. In the first case, an empty niche can be viewed as a "role" (sensu Elton) that no current species plays. This concept can also be seen in Odum's (1971) definition of the niche serving as the "address" of a species; such an address could be filled by any species with the appropriate traits. This type of situation might involve, for example, cases where a resource goes unused. In the second case, a given environment has a maximum potential diversity that represents the maximum number of niches that can be filled without causing extinctions. In resource competition, these two cases are linked; the maximum diversity is related to the number of resources (Levin 1970; Armstrong and McGehee 1980). Thus, new species might be able to invade during community assembly without causing any extinction until there is no resource base that is unutilized (i.e., one consumer for every resource) (but for examples where population oscillations can allow more species to coexist on fewer resources, see Armstrong and McGehee 1980; Huisman and Weissing 1999; Abrams and Holt 2002; see also chapter 6).

These issues become more complex when we consider the occurrence of multiple trophic levels. For example, interactions with species at higher trophic levels can create opportunities for coexistence of competing prey species. Grover (1994) has examined the assembly process for such food webs and shown that there is theoretically no limit to the diversity of the system. However, Grover (1994) also shows that the conditions for enhancing diversity become increasingly difficult as more species enter the community. More importantly, if the diversity achieved in the assembly process is highly dependent on the order in which species enter a community, then the concept of an empty niche becomes somewhat meaningless.

The implications of empty niches can be investigated by focusing on the distinction between the requirement and impact components of the niche. As we saw in the model of resource competition–environmental stress interactions (chapter 2), stress does not enhance local diversity because there is no feedback by which the species impact the level of stress. In cases where species have no impacts on the environment, questions about saturation or empty niches cannot be evaluated.

Thus, the idea of niches as roles that species play is much more closely related to the impact component than to the requirement component of the niche concept.

Niche versus Habitat

The concept of "habitat" is highly entangled with that of niche (Whittaker et al. 1973), principally because of the confusion as to whether niche refers to aspects of the environment or to aspects of the species. Habitat typically describes a subset of a species' niche, referring to the physical and some biotic features of the environment, but not explicitly interactions among other species. In fact, some authors have used the term "habitat niche" to describe this subset (Odum 1971; Grubb 1977; Smith and Smith 2000). Alternatively, habitats can simply be viewed as an area where an organism can live because it contains the conditions and resources that it needs. Our definition of the niche is explicitly mechanistic, and habitat represents a component of environmental supply that a species requires and impacts.

3.3. Concluding Remarks

Our revised niche framework seeks to clarify and resurrect the niche concept in a more quantifiable and biologically meaningful way than the previously loosely associated ideas of the niche. It has been our intent to clarify some of the ambiguities that arise when the niche is less formally defined, and also to provide an explicit link between our niche concept and earlier ones. The hundreds of papers published on more loosely defined niche concepts, as well as the specific formalization of Lotka-Volterra models of interspecific competition, had many important contributions. In order to develop a more synthetic ecology, it is important to make the connections between these classical and contemporary approaches more explicit and to expose both their commonalities and their differences.

3.4. Summary

1) For two species to coexist locally, per capita effects have to result in lower interspecific than intraspecific competition, the requirements of the species must show a trade-off, and the supply of a factor cannot disproportionately favor either species.
2) This basic conclusion is identical to that of the classic Lotka-Volterra models on which most previous niche theory is based.
3) The fundamental niche of an organism occurs where the supply point of two resources is above the ZNGI. The realized niche of

two consumers is narrower than the fundamental niche of either species, because they are competitively excluded from a portion of the range of resources.

4) In order to avoid confusion, niche axes must be quantifiable environmental factors, and several factors cannot be collapsed together into a single axis.
5) Niche overlap can determine only the short-term per capita impacts of interspecific competition.
6) The niche, as we have defined it, results from the interactions of the species within its environment. In this view, there is no such thing as an empty niche and species can't fill the same niche in different locations.

CHAPTER FOUR

DESIGNS AND LIMITATIONS OF EMPIRICAL APPROACHES TO THE NICHE

Until now, our treatment of the empirical side of the niche has been only cursory. We hope to spark empirical creativity, and so rather than dwelling too much on what has already been done, we will explore ways in which this type of approach will facilitate new insights. The approach that we advocate is not necessarily compatible with the way much of empirical ecology is currently done. This does not mean, however, that the niche framework is untestable. Rather, ecologists can greatly enhance their explanatory power by combining theory, small-scale experiments, and broad-scale comparative approaches in a niche-based framework.

Many of the concepts in the previous chapters have yet to be fully documented and tested. While the experimental approach to ecology that has dominated for the last twenty years has allowed us to verify or refute many ideas (Resetarits and Bernardo 1998), the way experiments are often designed does not allow us to fully view and understand the mechanisms underlying the observed patterns. For example, the factorial experiments that have become so commonplace explore how an effecter (such as a predator or competitor) influences some component of the responder (such as the density or growth of the species of interest) and analyze the results in analysis of variance (ANOVA) or some similar statistical test. These sorts of specific experiments tell us about the response to a particular treatment that leads to some pattern, but often don't say much about the underlying mechanisms or the generality of that pattern. Indeed, very few experimental data are available that allow a compelling look into the structure of the niche as we have defined it. Brown (1995) lamented the lack of comprehensive studies of the niche of any particular organism, and called for more detailed analyses akin to Connell's (1961a,b) classic studies of the biotic and abiotic limitations of intertidal barnacle species. Brown went so far as to suggest that Connell's decades-old research "still represents perhaps the most complete study of the niche of any species" (32). While Brown's call for the measurements of all aspects of the niches of "at least a few species" (31) seems a bit impractical, and probably unnecessary, his main point was that broader empirical studies exploring the niche are needed. We echo this call, but with a caveat.

It is probably quite impractical to measure every factor and component of a species' niche in the sense that Brown was discussing. However, it is practical to measure several factors, particularly in the sense of niche requirements and impacts, that are likely to be most important for a given species in a given locality (see also chapter 2). In the remainder of this chapter, we develop an overview of the ways one might use the niche concept as a framework for empirical studies. First, we discuss some ways in which specific details of the models have been tested previously, using examples from the laboratory and field. Second, we discuss some general ways in which the niche-based framework can serve as a basis for a variety of empirical research programs. Throughout, we develop the view that the utility of our models lies not with the direct testability or validity of all of their assumptions and predictions, but with their ability to serve as a springboard for a variety of important insights and syntheses. That is, we do not propose that this theory needs to be explicitly tested in the way empiricists often test models. Rather, we feel that the real value of theoretical models lies in structuring a conceptual view with which to develop and test qualitative hypotheses about the way a particular system is structured.

4.1. Quantitative Tests of the Niche Framework

We begin with a very brief overview of several quantitative tests of niche-based theory among real organisms. We do not intend the brevity of this section to downplay the importance of those results. First, there are only a handful of quantitative tests of the theory. Second, those tests have been well described elsewhere (see examples below). We emphasize that while quantitative tests of the conceptual framework are interesting and important, they are not necessarily required in order for the niche framework to be useful. This is because the framework can also be powerful as a hypothesis generator and in comparing broad (and thus necessarily coarse) ecological patterns.

Laboratory Experiments

Since Gause's (1936) classic experiments on the competitive exclusion principle, laboratory experiments have played an essential role in the ecologist's toolbox. Perhaps still some of the most carefully detailed work testing some of the basic tenets of the niche-based theory were performed on algae in chemostats by Tilman and colleagues (Tilman 1976, 1977; Tilman et al. 1981; Tilman and Sterner 1984; Tilman et al. 1986). Recall that the level of resources on which a consumer population does not grow or decline (zero net growth, at which birth rate = death rate) is termed that species' R^* for a given resource. When resources are essential (see chapter 5) rather than substitutable (a situation emphasized in chapter 2), in order for two species to coexist on two resources, each species must be a superior consumer (lower R^*) for one of the two resources. This is what Tilman found in competition experiments with algal species in laboratory chemostats (fig. 4.1). Furthermore, Tilman found that he could change the outcomes of competitive interactions among algal species by altering the relative ratios of the different resource types (Tilman 1977) or by changing the temperature of the microcosm, which altered the species' R^* (Tilman et al. 1981). Grover (1997) reviewed a number of other laboratory studies using the principles of a niche-based theory to describe the interactions among a variety of microorganisms (e.g., bacteria, algae, zooplankton) and found general agreement between the theory and observations. Indeed, the measured R^* of species correlated well with the outcomes of competition experiments in nearly every study. Similar types of laboratory experiments have been used to test concepts of resource competition and shared predation and have shown qualitative agreement with the predictions of the niche-based models (Bohannan and Lenski 2000; Steiner 2001) (fig. 4.2).

Laboratory research is a powerful way to test the basic ideas presented in the theoretical models. The number of laboratory experiments that have been carried out to test key ecological concepts has skyrocketed (e.g., Drake et al. 1996; Lawler 1998; Morin 1999). Experimentalists have tested a wide number of theoretical models using microcosm studies, and it is possible to test every module and type of situation discussed in this book and elsewhere (e.g., Grover 1988, 1990, 1991; Lenski 1988; Lawler and Morin 1993; Lawler 1993; Cochran-Stafira and von Ende 1998; Kaunzinger and Morin 1998; Fox 2002). However, as with any ecological tool, laboratory studies have their limitations (Carpenter 1996). For example, microcosm studies have been heralded as a way to verify the validity of the assumptions and predictions of theoretical models. When you think about it though,

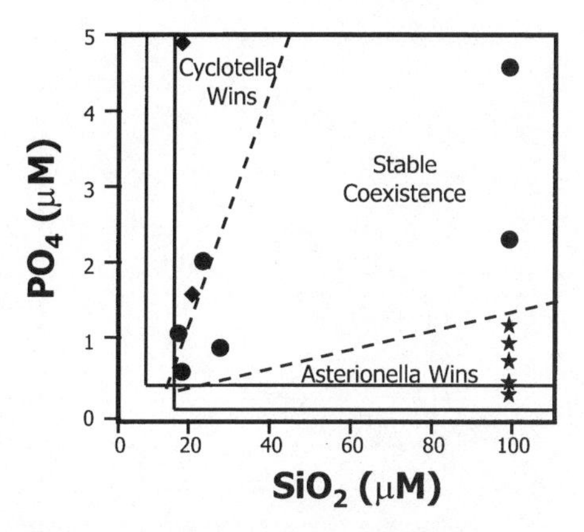

Fig. 4.1. Measured ZNGIs, impacts, and ranges of coexistence for two species of algae competing for essential resources in chemostats. Circles represent nutrient ratio treatments where the two species coexisted, stars represent treatments where *Asterionella* existed alone, and diamonds represent treatments where *Cyclotella* existed alone. Redrawn from Tilman (1977) with permission.

it should be possible to verify any mathematical theory should the appropriate species and environmental conditions be found. One could probably find species that will fit the assumptions of any model, but the outcome of such a laboratory experiment tells us very little about whether the assumptions of the models are meaningful in nature.

A Quantitative Field Test

Tilman and Wedin (Tilman and Wedin 1991a,b; Wedin and Tilman 1993) present what may be the only explicit quantitative assessment of simple niche-based models in a relatively natural situation. They measured the ability of several old-field plant species to draw down nitrogen from the soil when they were in monoculture (R^* for nitrogen), along with a variety of other traits (table 4.1) (Tilman and Wedin 1991b). They then used these values to predict competitive outcomes of pairwise competitive interactions as well as patterns of species composition along a nitrogen gradient, finding reasonably good concordance for most measures (Tilman and Wedin 1991b; Wedin and Tilman 1993). This study was somewhat limited in that: (1) R^* was measured on individuals planted in a monoculture rather than on a natural population of a species, (2) they estimated R^* within a single generation, which may not reflect the R^* at equilibrium, (3) they dealt with five species of grasses among the many other species of grasses and forbs that co-occur, (4) they dealt with only one limiting nutrient (nitrogen) among other potentially limiting nutrients and light, (5) they ignored herbivores, which are known to often influence com-

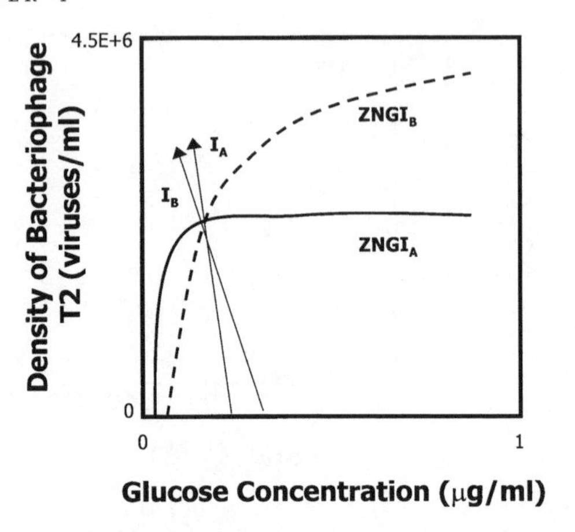

Fig. 4.2. Estimated ZNGIs, impacts, and ranges of coexistence from parameters based on a model system of two strains of *Escherichia coli* bacteria that consume glucose and are consumed by a bacteriophage. Redrawn from Bohannan and Lenski (2000) with permission.

petition among these plants (Ritchie et al. 1998), and (6) they did not measure the impact of these species on their environment.

The studies by Tilman and Wedin required a vast amount of treatments and measurements and the better part of five years of fieldwork, and yet it provides only limited insight into the mechanisms of coexistence among plant species in this community. So what's the point? Why bother espousing a framework that is supposed to be empirically friendly if it is so difficult to verify in nature? Do we need more and more studies that attempt to measure all of the details of requirements and impacts of species, which will undoubtedly result in a multitude of treatments and measurements? No, we do not believe that empirical studies based on the niche framework need be this complex and detailed. Instead, we advocate several ways by which to combine the theoretical predictions of niche-based theory with empirical observations and experiments in order to achieve broader insights without the time and finances necessary to explicitly measure the exact nature of the impacts and components of even a few species of interest.

4.2. The Niche in the Nature: Some General Considerations

We are concerned with three different types of empirical questions as they relate to this niche-based framework:

1) How can we test the general idea of the niche? Can we determine the most important limiting factors to a species? Can we show that species interactions and coexistence depend on understanding both

requirements and impacts? Can we show that a species' ability to invade a community is related to its relative requirements? Can we show that the stability of the species within a community is related to its relative impacts?

2) How can we test specific hypotheses using the niche-based framework? Can we test the predictions of specific models, such as the resource ratio competition model (e.g., Tilman 1982; Tilman and Pacala 1993) or the keystone predator model (Holt et al. 1994; Leibold 1996), in specific systems (e.g., algae in chemostats, algae in lakes, or algae in the presence of grazers)? Do we have an absolute standard for success (e.g., successful prediction within 5 percent of all quantitative predictions, or 20 percent, and so on) or a relative one (e.g., better fit than any previously specified alternative hypothesis)?
3) How can we use the niche-based approach to develop other testable ideas about species interactions? Presuming some degree of success, can we use the concept of the niche to generate testable new ideas, predictions, and applications? For example, can we use the niche concept and the ideas we have developed so far to study patterns of biodiversity? And do these ideas correspond to collected (or collectable) data on such patterns?

In this chapter we will limit our discussion to issues related to questions 1 and 2. This is because many of the examples that apply the niche concept to other aspects of species interactions (question 3) involve community, evolutionary, or ecosystem processes and patterns, which we discuss in later chapters.

We have already discussed the fact that even though it is possible to systematically search through factor after factor in pursuit of *the environmental axis* that might regulate the interaction between a pair of species, this is not the best way for an ecologist to spend their time. The empirical measurement of every single dimension of the Hutchinsonian niche will probably never be accomplished in a way that will be satisfactory to everyone. Even for a bacterium living in a chemostat, the entire *n*-dimensional hypervolume probably extends beyond the technical and practical abilities of any one researcher and is certainly not a very interesting question in and of itself. The issue of dimensionality shows that empiricists will have to be more pluralistic than simply going into the field and measuring traits or performing simple experiments.

In an attempt to present some structure to this creativity we have identified several important components of a successful empirical pro-

Table 4.1 Experimental results from Tilman and Wedin 1991b.

Species	R* (mg N kg soil^{-1})	Shoot N (% mass)	Root N (% mass)	Root: shoot mass
Andropogon gerardi	0.04	0.59	0.49	6.02
Schizachyrium scoparium	0.06	0.68	0.45	9.80
Poa pretensis	0.17	1.20	0.78	2.39
Agropyron repens	0.18	1.43	1.11	1.89
Agrostis scabra	0.33	1.17	1.24	0.85

gram centered on the niche framework. Often these studies start by identifying the most important limiting factors. They then calculate the responses to those factors as an estimate of its requirements and develop some conclusions about species' impacts on those factors. These studies are most successful when they combine those estimates to evaluate species' trade-offs. And they are most compelling and important when they combine theoretical models and cross-site comparisons to set these answers in a broader context.

Identifying the Relevant Limiting Factors

If we want to know the factors influencing the distribution of a particular organism, we first need to identify which factors are most likely to give us the most information—these factors will then become the axes of the requirements (ZNGIs) and impacts of the species. If we were studying a mouse, we would not likely consider oxygen as a limiting factor on which to draw a ZNGI. Even though oxygen is certainly one of the dimensions in the Hutchinsonian niche (the mouse can't live when oxygen levels are below a certain range), the mouse will probably never experience this limitation. Alternatively, oxygen can be a critical resource for which species compete if those species live in oxygen-poor environments, such as crayfish living in stream pools (Bovbjerg 1970).

Our graphical approach allows explicit consideration of usually two limiting factors at a time, even though extension to more limiting factors is possible (e.g., Huisman and Weissing 2001a,b). We feel that the benefits associated with the graphical framework often outweigh the benefits of considering more than two dimensions. In addition, as we discussed in chapter 2, two dimensions at a time are often all that is needed to gain considerable insights and predictive ability. For exam-

Table 4.1 (continued)

Species	Height (cm)	Light penetration (%)	Viable seed production (number m^2)	Maximum relative growth rate (wk^{-1})
Andropogon gerardi	10.00	93.00	6.00	0.21
Schiazachyrium scoparium	8.00	94.00	143.00	0.24
Poa pretensis	10.20	96.00	1030.00	0.34
Agropyron repens	13.60	92.00	11.00	0.13
Agrostis scabra	21.30	81.00	12500.00	0.41

ple, Liebig's (1840) law of the minimum posits that one (or a few) limiting factor(s) will determine the distribution of, and interactions among, species.

The important limiting factors in any given system are not always obvious and can depend critically on the type of question being asked. If we were interested in the types of traits a group of species possessed across an environmental gradient, then that gradient would be one of the niche factors. However, if we were to view the same system from a different perspective (say along a different type of gradient), we might use a different set of axes.

We can determine the factors that are limiting to a group of species of interest in a hierarchical way. As an example, we consider McPeek's (1996, 1998) detailed studies of freshwater larval damselflies. In many North American lakes, several species of damselflies appear to sort themselves out according to their abilities to compete for resources and deal with predators. Species of the genus *Ischnura* are uniformly superior resource competitors to those of the genus *Enallagma*, whereas those of the genus *Enallagma* are better at avoiding predators by having lower activity rates. The simple keystone predator module discussed in chapter 2 (e.g., Holt et al. 1994; Leibold 1996) can be used to determine the conditions (i.e., positions of the ZNGIs, impact vectors, and resource supplies) under which species within these genera should coexist or dominate (McPeek 1996). This answers a question at the genus level—how do the different genera coexist? Next, McPeek addressed a question at the species level—how do the different species within these genera coexist? McPeek showed that within the genus *Enallagma* (the genus less susceptible to predators), some species are better at living in lakes where dragonflies are the top predators, while other species are better at living in lakes where fish are the top predators (McPeek

1996, 1998). In order to explore the interactions and coexistence on this finer scale, it might be more appropriate to use these two types of predators as the limiting factors of interest, and draw ZNGIs in a manner similar to the apparent competition scenario discussed in chapter 2. Finally, even though this can help to explain different groups or guilds of species of *Enallagma* coexisting within different types of lakes, McPeek and Brown (2000) have still found there to be more species per habitat type than this scenario would predict. Thus, there may be still finer-scaled differences among these species, such as their abilities to deal with stochastic variance, habitat selection, or complexities due to their stage-structured life histories, all of which can be explored using a different set of axes that might help to explain this diversity. Alternatively, regional-scale patterns of metacommunity interactions could also affect the observed local scale diversity (see chapter 7). Finally, McPeek (personal communication) has argued that these species may be ecological equivalents whose coexistence may persist for long periods of time by a slow random walk to extinction (sensu Hubbell 2001).

Evaluating How Factors Affect the Fitness of Interacting Species: Estimating ZNGIs

Ideally, the niche-based framework requires measuring the ZNGI of each species along the relevant axes. This is not an easy task—very few ZNGIs have been measured in their entirety, and it is not likely that many ever will. Luckily, we don't have to measure the entire ZNGI in order to get a reasonable understanding of its shape. We discuss two empirical tools that could prove useful in evaluating the ZNGIs of organisms. The first is the "reaction norm" method that has gained popularity in both ecology and evolutionary biology (e.g., Schlichting and Pigliucci 1998; Tessier et al. 2000; Tessier and Woodruff 2002), which entails estimating components of fitness, such as growth rates and birth rates, in relation to the factor of interest. The second estimates the behavioral "decisions" of the organism of interest as an indicator of the costs and benefits in relation to the factor of interest (e.g., Brown 1988, 1989; Brown et al. 1994; Abramsky, Rosenzweig, and Pinshow 1991; Abramsky, Rosenzweig, and Subach 1997, 1998, 2000; Rosenzweig and Abramsky 1997).

A reaction norm is the response of the fitness of a species along a gradient of a factor. For example, if one wanted to discern how the fitness (or some indirect estimate of fitness) of a given species responds to a toxin, then the experiment would be simple—add more and more toxin to the system and measure what happens to the fitness of the species. Of course, we would expect the toxin to have a negative effect on the species, but by using several treatment levels of toxin, we could get some idea of the basic shape of the response. We can use the same approach to measure how organisms respond to many other types of factors that might be more relevant to explorations of the niche. Thus, by placing individuals of the species of interest in experiments where factors such as food availability, predation risk, or environmental conditions are varied, we can estimate the relative shapes and positions of the species' ZNGIs for these factors. For example, Tessier et al. (2000) have used this approach to explore how well different species of the freshwater zooplankton of the genus *Daphnia* grew on different densities of algal resources. By measuring how well individuals grew on a given level of resource, they could infer the relative R^* of each species; that is, the species able to have positive fitness on the lowest level of resource is assumed to have the lowest R^* (see also Garbutt and Zangerl 1983 for a discussion of this issue in relation to plants).

A second method for estimating the ZNGI of a species is to in a sense ask the species how it perceives the fitness of the environment around it. This method is based on the paradigm of optimality, which assumes that organisms can discern among habitats that provide differential fitness opportunities and make appropriate decisions. Optimality theory predicts that an individual foraging in a patch should continue foraging in that patch until the marginal value remaining in that patch is equal to the average value of the entire foraging area. This prediction, known as the marginal value theorem (Charnov 1976) can be used to predict how long a forager should remain in a patch before it gives up (known as its *giving up time*, or GUT). Building on this theorem, Brown developed (1988) and expounded upon (e.g., Brown 1989; Brown et al. 1994) a technique that uses the density of resources left in such a patch, rather than time, as the response variable. This *giving up density* (GUD) is the density of the resource in a given patch that is below the level where a reasonably judicious individual of a species wouldn't be caught dead (so to speak). Because R^* indicates the density of resources where a species' population has zero net growth and is a product of the traits of the species and its environment, and the GUD indicates the density of resources where an individual's fitness in a given patch is approximately zero, these two quantities are likely to be correlated. Thus, the GUD (and associated concepts) can provide an index of a species' relative competitive ability for a given resource and thus its R^* (Brown 1989; Brown et al. 1994; Chase et al. 2001).

In addition to measuring relative competitive ability, these behavioral techniques can also be used to estimate how varying levels of resources, predators, and other environmental factors influence a forager's ability to reduce resource levels (e.g., Abramsky, Rosenzweig, and Pinshow 1991; Abramsky, Rosenzweig, and Subach 1997, 1998, 2000; Rosenzweig and Abramsky 1997; Kotler and Brown 1999; Kotler et al. 2001), and thus they form a qualitative indirect estimate of the shape of a species' ZNGI. Furthermore, this tool is not irrelevant to plant ecologists. For example, Gersani et al. (2001) have shown how plants can allocate resources to differential root growth in relation to resource availability using concepts similar to GUDs (see also Kleijn and Van Groenendael 1999).

Evaluating How Species Influence Factors: Measuring Impacts

Some of the basic principles for measuring species' impacts are closely related to measuring food web interaction strengths. Measuring interaction strengths has been a popular activity among food web ecologists interested in the roles of species interactions within more complex communities. While the concept of interaction strength is itself somewhat complex (e.g., Laska and Wootton 1998; Berlow et al. 1999; Abrams 2001a), the basic question behind the concept is simple: what is the effect of a given species (such as a predator) on some factor (such as its prey)?

We can devise several relatively straightforward estimates of the impact of a species on a factor. For example, if one wanted to measure the impact of a plant on nutrients, we might add a variable amount of nutrients to the soil and measure the rate at which the plant is able to draw down those nutrients (e.g., Ryel and Caldwell 1998). Similarly, if we wanted to measure the impact of a plant on light, we would measure the availability of light above and below the plant and take the difference as its impact (e.g., Goldberg and Miller 1990). Of course, both of these measures would be complicated by how far the system is from equilibrium and the heterogeneity in nutrient and light availability induced by the plant itself. Alternatively, if we want to measure the impact of a predator on its prey, we could use behavioral observations—such as predation rates and predator density—similar to those espoused by Wootton (1997). The most appropriate method will depend on the particular system of interest. For example, measuring the impacts of a predator on its resource will be a very different task from measuring the impacts of a resource on its predator. In the former case we would measure some aspects of consumption rates, while in the latter we would estimate both the consumption rate of the predator and the conversion rate of the prey into new predators. Further, as we discuss in chapter 6, the relative shape of species' impacts can vary throughout the state space, and so the measurement of impacts may sometimes be sensitive to whether or not a system is near its equilibrium.

4.3. Evaluating Mechanisms of Coexistence and Trade-offs

Once an empirical program has evaluated the relevant limiting factors and obtained at least rough estimates of the requirement and impact components of a species' niche, it becomes relatively straightforward to examine how (or if) coexisting species differ in their attributes and whether trade-offs exist. We often assume that trade-offs allow species to coexist, but empirical verifications of this claim are essential to the development of a predictive theory based on the niche framework.

In the models described in chapter 2, there are usually two sets of linked trade-offs that must occur for species to locally coexist. Both of these trade-offs can be evaluated empirically using methodology like that described above. For example, Tessier et al. (2000) estimated the reaction norm of different species of *Daphnia* to estimate their R^* and then correlated this value with the ability of the species to uptake resources and grow. They found that species (or clones within species) that were better able to survive on low resources were worse at growing when resources were common (fig. 4.3a). Similarly, Chase et al. (2001) showed that several species of pond snails exhibited a trade-off in their ability to consume resources down to low levels within a patch and to find new patches (fig. 4.3b).

Once the ZNGIs and impacts are estimated and relevant trade-offs observed, we can begin to make synthetic predictions. In the traditional experimental approach, an ecologist will often remove or add species in combination with other factors and observe the outcomes. This represents a phenomenological approach, usually stemming from questions such as "do these two species compete?" and "do herbivores matter for understanding plant abundance and diversity?" However, armed with the predictions that one gains from a niche-based approach, experiments can be done in a much more deductive way. Here, we would instead ask questions such as "when do we expect competition to matter, and when don't we?" and "under what conditions should herbivory be more important?" We believe that this sort of approach, particularly when intimately linked with a theoretical foundation, will lead to a much stronger foundation for a synthetic approach to doing ecology.

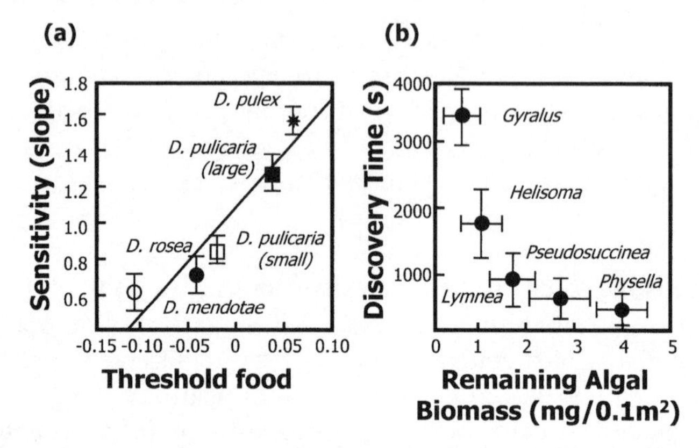

Fig. 4.3. Two empirical approaches to measuring trade-offs among species. (a) Trade-offs among *Daphnia* species (and clones) in their relative ability at surviving on resources when resources are rare and to consume resources when resources are abundant. Redrawn from Tessier et al. (2000) with permission. (b) Trade-offs among pond snail species (only genera names are given) in their relative ability to consume resources to low densities within patches and their ability to find new patches. Redrawn from Chase et al. (2001) with permission.

4.4. Concluding Remarks

There are several empirical research programs that fit well into the framework that we have been discussing, and although they do not measure all aspects of niche requirements and impacts, they have come close. Examples from very different types of systems include:

(1) detailed explorations into the mechanisms of plant competition and the effects of resources and resource ratios on that competition (Tilman 1982, 1988; Wilson and Tilman 1995; Inouye and Tilman 1995), as well as the modifying effects of herbivory (Ritchie and Tilman 1995; Ritchie et al. 1998), disturbance (Wilson and Tilman 1993; Tilman 1994), and dispersal limitation (Tilman 1997);
(2) studies of interactions among desert rodent communities, which explored the role of competitors (Abramsky, Rosenzweig, and Pinshow 1991; Abramsky, Ovadia, and Rosenzweig 1994; Abramsky, Rosenzweig, and Subach 2000; Brown et al. 1994; Rosenzweig and Abramsky 1997) and predators (Kotler 1984; Abramsky et al. 1997, 1998; Kotler et al. 2002) in restricting the habitat use and interspecific interactions among species;
(3) explorations into the mechanisms of competition among coexisting sunfishes (Werner and Hall 1976, 1977, 1979) and the modifications induced by predation (Werner et al. 1983) and size-structure (Werner and Hall 1988; Osenberg et al. 1992).

Of course, these are merely some of the many exemplary empirical research programs that utilize concepts similar to the ones we espouse. These research programs converged in a number of important aspects. First, each program identified the relevant mechanisms of species interactions and let the important biology of the system determine the next directions of research. Rather than finding a theory or asking a particular question, and then finding the "model" system that best fits the assumptions and hypotheses of that question, a niche-based approach allows the system to lead to the important questions to ask. While each system has its own quirks, the niche-based empirical framework provides a common approach that links these otherwise disparate investigations.

Often, empirical researches seek to test models with a particular system and accept or refute that model based upon their results. Similarly, theorists often look for empirical research to directly test the predictions and assumptions of their models. In both of these cases, there is little direct interplay between theory and empirical work. The niche-based framework makes the interplay between theory and empirical work much more direct. Niche-based theory is flexible and intended to provide empirically *friendly* qualitative predictions that can be compared and contrasted in a wide variety of systems. Thus, it provides a powerful conceptual tool that should be of great use in the study of most natural systems. Furthermore, when appropriate complexities are discovered in a particular type of system, many of these complexities can be incorporated. In the next chapters, we show some ways to do this.

4.5. Summary

1) The key features of a successful research program using a niche-based framework are: (a) identifying the question and the important limiting factors, (b) identifying the relative positions and trade-offs in species requirements and impacts through observations, inferences, and experiments, (c) using these insights to make theoretical predictions, and (d) testing these predictions with observations and experiments.
2) We suggest that empirical measurements of the entire niche of a species are difficult and unwarranted. Instead, creativity in the context, design, and analysis of ecological observations and experiments will go a long way in determining the key features of a species' niche and its interspecific interactions.
3) Methods already developed to measure niche requirements in-

clude measuring reaction norms across environments and foraging choices (GUDs).

4) Measurements of interaction strengths are analogous to measuring niche impacts.
5) We suggest that the most successful empirical research programs combine theoretical predictions, short- and long-term experimentation, cross-scale comparisons, and a healthy knowledge of the natural history of a particular system.

CHAPTER FIVE
INCORPORATING BIOLOGICAL COMPLEXITIES

So far, we have described models that make some simplifying assumptions about species interactions. In the next two chapters, we will add more biological realism to the framework, which sometimes causes interesting deviations from the predictions of simpler models. While many forms of biological complexity are possible, this chapter focuses on four types: resource utilization (how interacting species differentially use distinct resource types); behavioral plasticity (variation in a species' behavioral traits in response to environmental, including biotic, conditions); resource allocation plasticity (variation in a species' morphological or physiological traits in response to environmental, including biotic, conditions); and demographic structure (variation in stage, size, and other factors that change with age in a species). Chapter 6 discusses the influence of environmental variability and nonequilibrial dynamics. In each case, we discuss how the particular complexity influences the shape of the ZNGIs, the impact vectors, or both, and thus how it alters the outcomes of interspecific interactions.

5.1. Interspecific Competition among Species with Different Types of Resource Utilization

Generalists versus Specialists

We have assumed that two competing species share all types of resources, but complete resource overlap is rare. For example, consider a case where one species treats two resources as substitutable (a generalist), while another consumes only one resource type (a specialist). The ZNGIs for these two species will intersect provided that the specialist is able to maintain a lower R^* for its resource than the

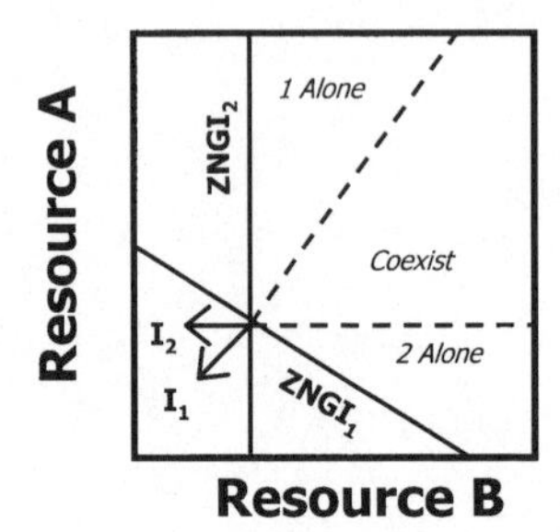

Fig. 5.1. Interactions between a generalist ($ZNGI_1$) and a specialist ($ZNGI_2$) for two resources. The ZNGIs intersect so long as the specialist is a better competitor for the resource on which it specializes. Impacts and ranges of coexistence are as in previous chapters.

generalist (fig. 5.1). That is, the species trade off such that the specialist is a better competitor for its narrower range of resources, while the generalist is a relatively worse competitor but can coexist by consuming other types of resources (see also Hutchinson 1957; Miller 1967; Colwell and Fuentes 1975; MacNally 1995; Chase 1996a,b).

The generalist-specialist trade-off alone, however, does not guarantee coexistence, as the supply point of the two resources must also be taken into account. Provided the ZNGIs intersect, the equilibrium will always be stable, and there will always be a range of resource supplies where the two species can coexist locally, as the specialist has no impact on one of the resources. However, if the generalist's exclusive resource (resource A) is in relatively high abundance, the generalist will outcompete the specialist. Alternatively, if the supply of the generalist's exclusive resource is relatively low, the specialist will outcompete the generalist. Only when the supply point falls within an intermediate ratio of the two resource types will the generalist and specialist coexist (for an empirical example of this phenomenon, see Chase 1996b).

Over thirty years ago, MacArthur (1972) considered a similar scenario, but with two specialists and one generalist. If the two specialists' ZNGIs intersect below the intersection of the generalist's ZNGI with that of either of the two specialists (each of which specializes in a different resource), the two specialists can coexist and competitively eliminate the generalist, so long as the resource supply exceeds the minimum requirements for each specialist (fig. 5.2a). That is, the generalist will be competitively excluded if the two specialists, at their intersection, can collectively reduce both resources down to levels lower than that at which the generalist can survive. Alternatively, if the generalist's ZNGI intersects with both specialists' ZNGIs at a point below the intersection of the two specialists' ZNGIs, then a richer set of outcomes is possible. If the resource supply is strongly skewed toward one specialist's resource (i.e., below the other specialist's ZNGI), then

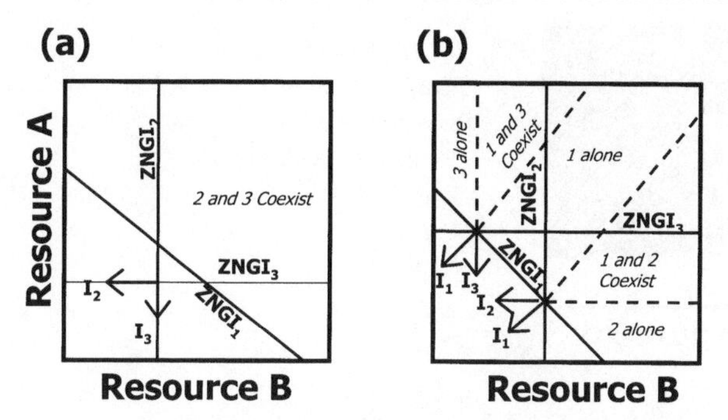

Fig. 5.2. A generalist ($ZNGI_1$) with two specialists ($ZNGI_2$ and $ZNGI_3$). (a) The intersection of the $ZNGI_2$ and $ZNGI_3$ occurs below $ZNGI_1$ and the two specialists coexist but exclude the generalist. (b) The intersection of $ZNGI_2$ and $ZNGI_3$ occurs above $ZNGI_1$, and the outcome becomes more complicated.

that specialist will be able to outcompete the generalist and exist alone. If the resource supply is only slightly skewed toward one specialist's resource, that specialist will coexist with the generalist. Finally, if the supply of the two resources is intermediate, then the generalist will be able to outcompete both specialists and exist alone (fig. 5.2b). Thus, this simple scenario can lend insight into the environmental conditions (i.e., resource supply ratios) in which specialization or generalization would be more favored (see, e.g., Brown 1996).

Nonlinear Resource Utilizations

In chapter 2, we focused on substitutable resources, which yield linear ZNGIs. There are many physiological reasons, however, to suspect other types of resource utilizations would lead to differently shaped nonlinear ZNGIs. For example, the best-known graphical models of resource competition focus on essential resources, which lead to an L-shaped ZNGI, since each species requires a minimum amount of each resource type (fig. 5.3). When resources are essential, the criteria for coexistence are qualitatively similar to those discussed for substitutable resources in chapter 2 (see also Tilman 1982, 1988; Grover 1997).

In cases where the ZNGIs of two species differ significantly in their shape, then those ZNGIs may intersect at two different points and lead to fundamentally different outcomes. Fig. 5.4 presents a simple example of this phenomenon when one species has a linear ZNGI and treats resources as substitutable, while the other has a nonlinear ZNGI and treats the same resources as essential. (We were unable to think of a realistic empirical example to match this scenario, but we use this special case to illustrate the general phenomenon of multiply intersecting

Fig. 5.3. The ZNGI (dark line) for a species that treats two resources as essential. The shaded area above the ZNGI represents the requirement component of this species' niche, where birth rates are greater than death rates.

ZNGIs.) When ZNGIs intersect twice, there are two potential equilibria. To evaluate the outcomes of such interactions, we need to consider the impact vectors at each intersection. In the upper left-hand portion of fig. 5.4, the species that treats the resources as essential (species 2) is more limited by resource B than the species that treats the resources as substitutable (species 1). In the lower right-hand portion, species 2 is more limited by resource A than species 1. That is, the resource that species 2 finds most limiting switches between the two equilibria. However, the relative position of the impact vectors does not switch unless the organism has different consumption strategies with different resource supplies. Thus, regardless of which resource species 2 most greatly impacts, it can have a greater impact on the resource that least limits it at one intersection and on the resource that most limits it at the other intersection. This will then lead to one ZNGI intersection being a stable equilibrium and the other unstable. In addition, we explore the range of supply points where these equilibria will be expressed by extending the inverse of the impact vectors and find that there will be a range of resource supply where the inverse of the impact vectors for the ZNGI intersections overlaps. When the resource supply falls within this overlap, which species invades the habitat first will determine which of the ZNGI intersections will be realized and which final outcome will be expressed. Tilman (1982) discussed a variety of other situations of resource utilization that could lead to similar patterns of multiply intersecting ZNGIs but assumed that the relative position of the two species' impacts could shift between the two equilibria, leading to a qualitatively simpler prediction. Nevertheless, increasing the complexity of the resource utilization among species can lead to complex patterns regarding the interactions among competitors.

5.2. *Behavioral Plasticity*

Original studies on foraging theory and habitat selection in the context of behavioral ecology by MacArthur and Pianka (1966) were devel-

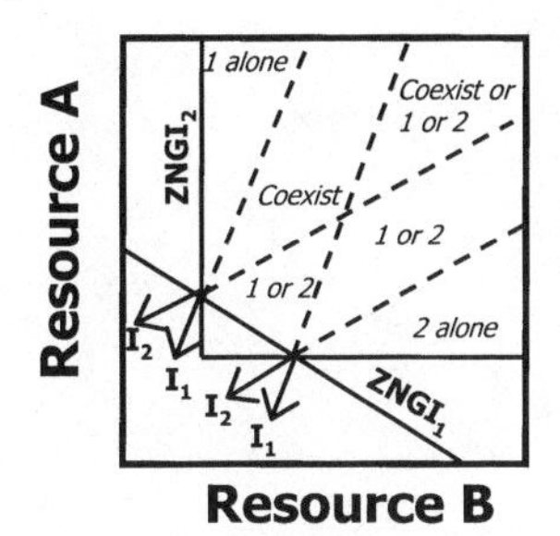

Fig. 5.4. Interactions among two species for two resources (A and B), where one species treats resources as substitutable, and the other species treats resources as essential. When the ZNGIs intersect in two places, several outcomes are possible, depending on the level of resource supply, the relative position of the impact vectors, and in some cases, the order in which the species enter the community.

oped explicitly in conjunction with MacArthur and Levins's (1967) model of niche relations (see chapter 1). In fact, behavior is an implicit component of many of the parameters that underlie the graphical models of ZNGIs and impacts. For example, the resource consumption rate (functional response) and susceptibility of prey to predators (through behavioral avoidance) are implicit parameters in the analytical models presented in the appendix to chapter 2. However, if behavior explicitly influences the shape of the requirement and impact components of a species' niche, then it can alter the predicted outcomes of interactions. Such emergent effects of species behavior on niche dynamics and community interactions have been an area of fruitful research in recent years (see Fryxell and Lundberg 1997). In this section, we focus especially on foraging behavior and patch choice to illustrate cases where allowing explicit behavior can significantly alter our predictions within a niche-based framework.

Habitat Selection without Travel Costs

Consider a species that has two substitutable resources, each resource occurring in a distinct habitat, assuming, for now, that there are no costs associated with a forager moving between patches (see Vincent et al. 1996). Thus, the forager cannot consume both resources without moving among habitats. Here, we assume the consumers will choose habitats that maximize their individual fitness. The consequence of this behavior is that at equilibrium, the consumers distribute themselves in the two habitats such that fitness among individuals is equivalent. When there are two habitat types separated in space, an individual that is foraging in one habitat type can not forage in the other. To determine in which habitat type an optimal habitat selector would choose to forage, we can derive an isofitness line. This line indicates the availability of resources where an individual has equal fitness in either habitat type (fig. 5.5). If the supply of the habitat types is above or below this line, then individuals should prefer one or the other habitat type. In habitat

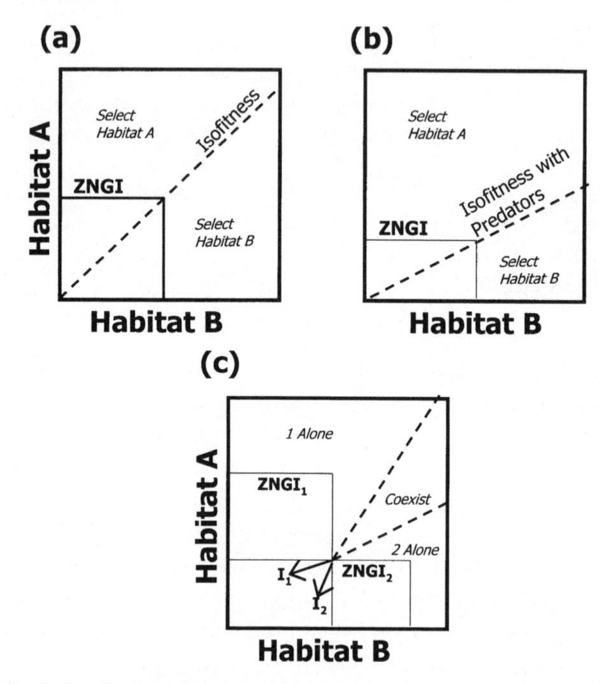

Fig. 5.5. ZNGIs of a species that selects between two habitat types. (a) The isofitness line, denoting the position of equal fitness between the two habitat types; the vertex of the ZNGI is on this line. (b) The effects of predators in one habitat on the position of the isofitness line and ZNGI. (c) The ZNGIs and isofitness lines of two species that use resources in two spatially segregated habitats but differ in their relative ability to use those resources.

selection models, consumers cannot substitute habitat types, even though the resources themselves are substitutable, and must choose to be in any one habitat or the other (Vincent et al. 1996). To derive the ZNGIs for this situation, we draw a straight line (vertical or horizontal) from the minimum resource requirement of the species (R^*) in each habitat to the isofitness line (fig. 5.5a).

An interesting extension of this model occurs when the two habitats differ not only in resources, but also in some other fitness component, such as predation risk. Under simplifying assumptions, if a forager is faced with two habitats that differ in their predation risk, an optimal forager should maximize the ratio of growth to mortality in its foraging decisions (e.g., Werner and Gilliam 1984; Gilliam and Fraser 1987). Increasing the fitness costs in one habitat due to predation risk will shift the isofitness line toward the less risky habitat, and the position of the ZNGI will shift (fig. 5.5b). In fact, a similar result will be obtained any time the costs or benefits unrelated to foraging gain itself differ among the patch types. Adding more realistic constraints, such as diminishing returns of fitness gain from food, will round the upper corner or the ZNGI, but not alter the qualitative predictions (J. S. Brown, personal communication).

If each habitat type contains a different essential resource, then spatial segregation of resources will have a less dramatic effect on the shape of the ZNGI. This is because both resources are essential for the fitness of the species. In this case, the species will require a minimum amount of each resource, and habitat selection will round the corners of the ZNGI (since the forager cannot use both resources simultaneously) from the distinct L shape that is seen when resources are not segregated spatially (see Vincent et al. 1996).

Although habitat choice changes the relative shapes of the ZNGIs when resources are present in distinct habitats, the local coexistence criteria will not differ qualitatively from the case when resources are spatially aggregated. That is, each species must differentially use the two habitat types so that their ZNGIs intersect, they must have the strongest impact on the resource type they find most limiting, and the ratio of resource (and habitat) supply must be intermediate (fig. 5.5c).

Scales of Perception

In some cases, two species that consume the same resources can differ in their ability to perceive those resources, which can dramatically affect the requirement and impact components of their niches (Ritchie and Olff 1999). A forager may see the two resource types as if they were in the same habitat patch and would respond to those resources without making habitat choice decisions. Such a species is sometimes said to be a fine-grained forager. Alternatively, a coarse-grained forager perceives those same resources as being in separate habitat patches (sensu MacArthur and Levins 1967; Rosenzweig 1981). For example, a grasshopper that consumes two different types of plants might view those plants as occurring in different habitat types because it must travel from one plant to the other to consume them, while a bison consuming the same plant resources would perceive them as occurring in the same habitat and would simply consume some proportion of each plant type—perhaps even in a single bite! However, even similar-sized species can treat the distribution of their resources differently (Ritchie and Olff 1999).

If species with these two different foraging modes compete for the same resources, the relative shapes of their ZNGIs will differ and create some interesting predictions. Two species' ZNGIs will intersect, or trade off, once if the species that is a superior competitor for one resource

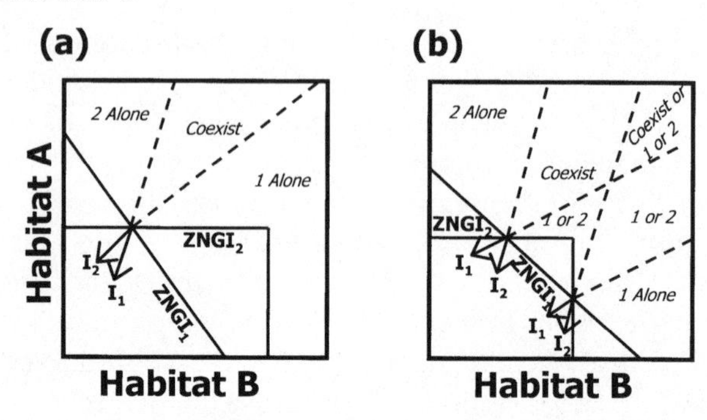

Fig. 5.6. ZNGIs for two species that perceive resources on different scales. Species 1 (a coarse-grained forager) treats resources as if they were in the same habitat; species 2 (a fine-grained forager) treats resources as if they were in different habitats. (a) Each species is superior at consuming a different resource type. Coexistence criteria are similar to previous discussions. (b) The fine-grained forager is worse at consuming both resources, but its ZNGI intersects that of the coarse-grained forager in two places. As in fig. 5.4 several outcomes are possible, depending on the level of resource supply, the relative position of the impact vectors, and, in some cases, the order in which the species enter the community.

is an inferior competitor on the other (this applies both to substitutable and to essential resources) (fig. 5.6a). However, if the two species are relatively similar in their resource requirements (i.e., if one species has lower ZNGIs for both resources), but they differ in their scale of perception, then the ZNGIs of the two species can potentially intersect in two separate locations (fig. 5.6b). In this case, then, coexistence can occur without a trade-off in the ability of the species to utilize the two resources because of the differences in the way they perceive the habitat. In addition, the likelihood of coexistence will also depend on the levels of resource supply, some of which will lead to unstable equilibria, in a manner qualitatively similar to what we discussed above when species utilized resources differentially (fig. 5.4).

Habitat Selection with Travel Costs

We briefly consider the case where we add travel costs to the simple model of habitat selection for substitutable resources (see Brown 1992, 1999; Leibold and Tessier 1997). Costs occur when individuals lose opportunity, lose energy, or encounter mortality risks (e.g., predation) when traveling. When travel costs exist, there are two isofitness lines representing two behavioral thresholds, each one determining whether to stay in each patch type when it is encountered (fig. 5.7a). Between these lines, the species should use patches of both habitats in

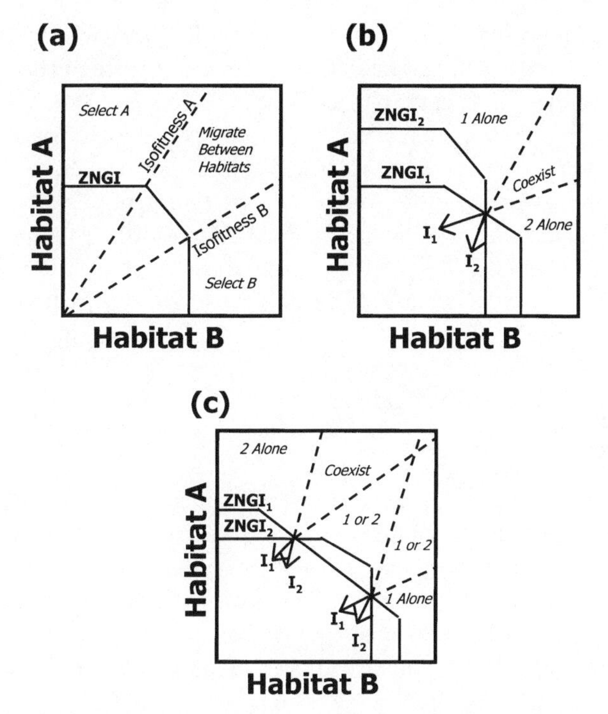

Fig. 5.7. ZNGIs for species that choose between two habitats and have costs associated with moving between them. (a) The ZNGI of a species that has thresholds where it will remain in one habitat or the other, and a range where it will migrate between the two habitats substitutably that are determined by the position of the isofitness lines. (b) ZNGIs of two species intersect once, showing typical coexistence criteria. (c) ZNGIs intersect twice due to differential costs of migration. As in fig. 5.4, several outcomes are possible, depending on the level of resource supply, the relative position of the impact vectors, and, in some cases, the order in which the species enter the community.

a substitutable way. The ZNGIs for this relationship are similar to those that exist when there are no travel costs, except in this case the two orthogonal lines from the axes are connected by a line representing the range of conditions where the benefits of migrating outweigh the travel costs, and individuals use both patch types.

In this case, two competing species with similar traits will coexist if their ZNGIs intersect, and their impact vectors are positioned so that each species has a stronger impact on the resource (habitat) by which it is most limited (fig. 5.7b). However, where one species has large travel costs between patches while another has very small (or no) travel costs, it is possible that the species' ZNGIs can intersect at more than one location, such that the coexistence and abundance of these species depend on both supply rates and initial conditions (fig. 5.7c).

These models may help us understand and predict why species will remain within areas or migrate at considerable costs between different types of habitats. For example, Leibold and Tessier (1997) used this modeling framework to predict how two species of *Daphnia* in lakes choose either to remain within a single habitat (i.e., the top epilimnion layer or the lower hypolimnion layer) or to migrate between habitats on a daily basis (known as diel vertical migration). This model was based on the benefits of food resources gained from each habitat combined with the costs of migrating between habitats, particularly due to predation by fish. Similar models might also be useful in understanding the long-distance migratory habits of a variety of animal species (e.g., Neotropical birds, wildebeest).

5.3. Resource Allocation

In this section, we illustrate some ways to incorporate into the niche framework how species adapt to local environmental conditions by shifting allocation of resources to different traits. We use the examples of plants allocating resources to roots or to leaves and prey allocating resources to growth or defense against predators. The framework can be adapted to the myriad other forms of such plasticity in similar ways.

Allocation to Roots versus Leaves

Consider a plant that can allocate its biomass to either root tissue or leaf and shoot tissue. Allocation to roots facilitates the plant's ability to acquire underground resources such as water and nutrients, whereas allocation to shoots and leaves facilitates the plant's ability to acquire light. In the short term, if an individual allocates resources to roots, it cannot reverse this on a whim and switch those resources to shoots and leaves (though some perennial plants can shift allocation strategies over time) (e.g., Tilman 1988; Gleeson and Tilman 1992; Reynolds and Pacala 1993; Rees and Bergelson 1997).

One can visualize the effect of plasticity in resource allocation by viewing each allocation strategy independently along the niche axis of interest. Because many plants are plastic in their ability to allocate resources to roots and leaves, each allocation strategy can be thought of as having a distinct ZNGI (light lines in fig. 5.8a). The overall ZNGI for the species with allocation plasticity is then represented by the lowest points of each allocation strategy (which for something like resource allocation can be approximately continuous), and is concave up, because the different resource types are essential resources (i.e., underground resources [nutrients and water] and aboveground resources [light]) (dark line in fig. 5.8a).

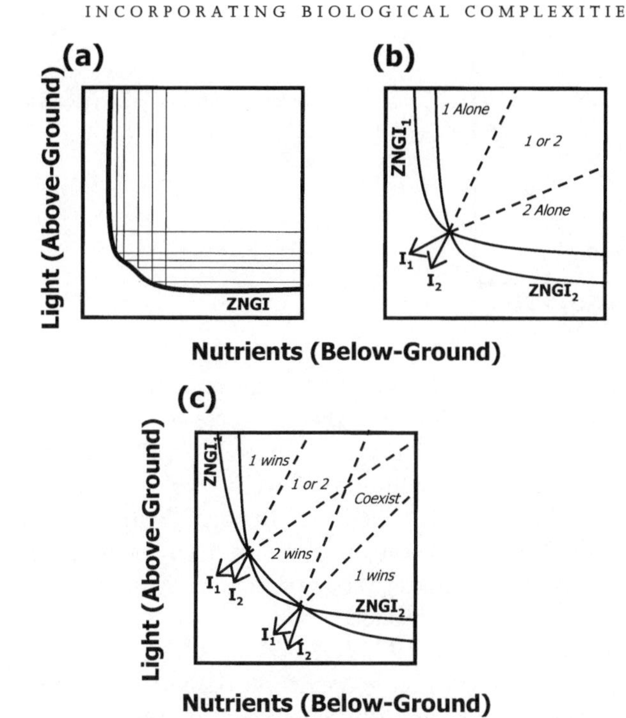

Fig. 5.8. ZNGIs for species that are able to differentially allocate resources to aboveground and belowground structures. (a) Derivation of the overall ZNGI. Each light line represents a ZNGI for a discrete allocation strategy (allocation to roots or shoots), whereas the dark line represents the overall ZNGI for the species. (b) ZNGIs and impacts of two species with differential allocation to belowground and aboveground resources. Note that the position of their impact vectors I is such that the equilibrium is unstable, and the species that enters the community first is expected to dominate and outcompete the other at intermediate supply ratios of the resources. (c) A case with two species that have different allocational abilities; thus, their ZNGIs intersect twice. Several outcomes are possible, depending on the level of resource supply, the relative position of the impact vectors, and, in some cases, the order in which the species enter the community.

Now consider the case where two species are competing for light and belowground resources. If both species exhibit similar levels of plasticity, then their ZNGIs will intersect at one point provided they trade off such that one species has a lower R* for one resource (i.e., it is a better competitor for that resource type), while the other has a lower R* for the other resource type. Because of the location of soil resources (belowground) and light resources (aboveground), the relationship between requirements and impacts will differ from what we have generally assumed. Plants that allocate more to belowground resource acquisition (roots) will not only have a lower ZNGI for that resource but will also have greater impacts on that resource. In addi-

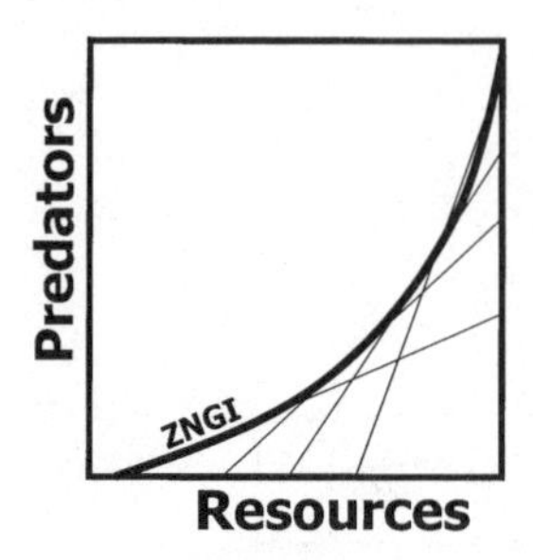

Fig. 5.9. A species with the ability to differentially allocate resources to reproduction or defense against predators. Thin lines are the separate ZNGIs of each discrete strategy, and the thick curve is the overall ZNGI.

tion, plants that allocate more to aboveground resource acquisition (leaves) have a lower ZNGI for that resource and also have greater impacts on that resource. Thus, even though the ZNGIs of the plants intersect, the equilibrium will not be locally stable. Instead, the initial conditions of the system will determine the final "winner" of the interaction (fig. 5.8b). The predictions we derived using these simple graphical tricks are qualitatively comparable to those derived analytically and with simulation models (Tilman 1988; Reynolds and Pacala 1993; Rees and Bergelson 1997).

Finally, if two species differ in the degree of their allocational plasticity, their ZNGIs can intersect at two points, indicating the potential for two alternative equilibria (fig. 5.8c). In this case, the relative positions of the impact vectors are as above, and so the final outcomes in this system are determined by a complex combination of supply points and initial conditions (see Gleeson and Tilman 1992).

Allocation to Growth or Defense against Predators

Some prey species can morphologically or chemically defend themselves against predators when predation risk is high, but allocate resources to other functions when the risk is low (the same can be said for plants with induced defenses) (Tollrian and Harvell 1998). We assume that a prey species can differentially allocate resources to growth and/or fecundity or to traits that confer a defense against predators. Thus, allocation of resources to defense is assumed to incur a fitness cost. Here, we again determine the overall shape of a species' ZNGI by first drawing the ZNGIs of each separate strategy, assuming their differential costs and benefits (thin lines in fig. 5.9), and then by taking the outer boundary of these as the overall ZNGI of the species (thick curve in fig. 5.9). Chase (1999b) showed an analytical solution to this problem yielding identically shaped ZNGIs.

This simple scenario can allow us to make some predictions about the consequences of plasticity to the interactions and dynamics within the community. For example, as resource supply increases, the relative ability of predators to control prey decreases. This is because the overall slope of the ZNGI increases with increasing productivity, and thus prey are proportionately influenced less as resource supply increases (fig. 5.9). Biologically, this result occurs because at high levels of resources, the benefits of defense become increasingly higher than its costs, and prey are expected to exhibit defensive strategies more readily, since the ZNGI for the defended strategy is above that for the undefended strategy (see also Chase, Leibold, and Simms 2000). Although we do not belabor the main point, one could imagine a variety of scenarios where differential allocational strategies to growth or defense among species could lead to a variety of outcomes, including multiply intersecting ZNGIs. We return to the subject of plasticity in this sense in chapter 10, when discussing how these traits might evolve, and what environmental conditions would favor or disfavor plasticity in the allocation to defensive traits.

5.4. Population Demographic Structure

Often, organisms change considerably throughout their life cycle. In fact, life stage transitions can be so extreme that a species occupies functionally different niches at different times; this scenario is sometimes referred to as an "ontogenetic niche shift" (e.g., Werner and Gilliam 1984). A huge diversity of organisms undergo such ontogenetic niche shifts, including holometabolous insects and amphibians, which undergo complete metamorphosis from larval to adult forms, and fishes and reptiles that can completely shift diet as they grow. Less dramatic but not less important changes in life cycle occur in many other groups of organisms, such as plants that grow through many distinct stages (e.g., seedling, sapling, and adult trees). Studies have shown that the outcomes of species interactions and coexistence often result from, and are complicated by, variation in population structure (e.g., Werner 1986; Ebenman and Persson 1988; Osenberg et al. 1992; McPeek and Peckarsky 1998; Persson et al. 1999; Chase 1999a,b; Knight 2002).

In this section, we outline a few ways to incorporate population structure and ontogenetic changes into the niche framework. We explore: (1) the size-structured interactions between juveniles and adults that compete for similar resources, (2) the ontogenetic shifts of a species that goes through larval and adult stages in different habitats, and (3) the effects of size-structured predation on species competing for limiting resources.

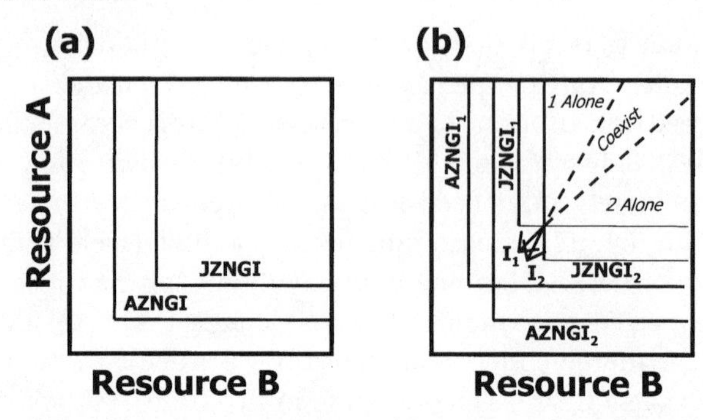

Fig. 5.10. A way to depict species' distinct life stages in the graphical framework. We assume that a plant has two distinct stages, adult and juvenile, and utilizes two essential resources. We draw separate requirements for adults (AZNGI), which depict the level of resources needed to balance juvenile transition to adults and adult death, and for juveniles (JZNGI), which depict the level of resources needed to balance birth rates with juvenile transition to adults. (a) A case where adults are uniformly better at consuming resources than are juveniles. (b) Competition among two species with two life stages. The interaction of the JZNGIs only determines the possible equilibrium of the system, but the impacts are the joint effects of both adults and juveniles.

Size-structured Competitive Interactions

Consider a case where a species has considerable size structure that influences its ability to uptake and convert resources. We first assume that adults are uniformly better at consuming resources than are juveniles. For example, seedling plants will likely be much worse at consuming and converting soil resources and water (due to small, underdeveloped root systems) as well as at capturing light (due to short stature and underdeveloped leaf systems) than will adults of the same species. Thus, it is the requirements of the juvenile that will determine whether or not it will become an adult in the first place. That is, in a spatially homogeneous environment or when a species is sessile (such as a plant) it does not matter if the requirements and impacts of the adult favor a species' existence or even dominance if the juveniles cannot thrive in that environment. This implies a very close link with demographic models of population ecology by linking the definition of ZNGIs of species with complex life cycles to the conditions that allow "closing" of the life cycle (e.g., Caswell 2001).

A simple trick that we use to incorporate size, stage, and age structure into the niche framework is to draw separate requirements and impacts for each grouping. Thus, for an organism with juvenile and adult life stages, we can estimate a requirement component for each stage (fig. 5.10a). First, there is a minimum amount of resource necessary for a juvenile to have a zero net transition rate into and out of the juvenile stage (JZNGI). If fewer individuals transition out of the juvenile to the adult stage than the birth rate of juveniles from the adult stage, the species cannot persist in that environment. Second, there is a minimum amount of resource necessary for an adult to have a zero net balance between juveniles transitioning into the adult stage and the death rate of adults (AZNGI). In this example, the species can persist in an environment only when the resource supply exceeds the JZNGI.

Now, if we consider the interspecific competitive interactions between two species that both have structured populations, we first need to consider the positions of the JZNGIs, which for now we assume are above the AZNGIs. If the JZNGIs do not intersect, the species will not coexist, regardless of whether or not the AZNGIs intersect. Instead, the species with the lower JZNGI should win and exist alone. If the JZNGIs do intersect, then coexistence is possible but becomes somewhat more complicated. First, the supply must lie above the JZNGIs. Second, the impact vectors combine both juvenile and adult impacts on the same resources (fig. 5.10b). If the juveniles of one species have a stronger impact on both resources, we would predict this species would exist alone if we were only considering juveniles. However, the relative impacts of the adults could reverse this outcome and allow coexistence or even dominance by the other species.

So far, we have assumed that adults can modify the outcome of competitive interactions only through their impacts. However, adults can modify competitive outcomes through maternal effects (i.e., resources supplied to offspring from the mother, such as seed size). In essence, maternal effects make it easier for juveniles to transition into the adult stage, because they allow juvenile plants to grow through early life stages as a result of the adult plant's resources, not the juvenile's, and lower the relative position of the JZNGI.

Fig. 5.11 shows another feasible situation, where juveniles and adults consume and compete for two resources, but juveniles are better at consuming one resource type and adults at another. For example, many types of species, such as turtles or lizards, are omnivorous. However, smaller individuals (i.e., juveniles) are typically worse at consuming plant material because they have shorter guts that do not allow efficient digestion of plant material, but better at consuming animal material than larger individuals (i.e., adults). In such a case, we can draw the ZNGIs of the two stages as above, but here they will intersect, and the overall ZNGI of the species represents the minimum resource requirements of each stage. Note that each stage constrains the other in its ability to survive on low resources. From this, averaging the im-

Fig. 5.11. A species with two distinct stages, adult and juvenile, which require two limiting resources. In this case, juveniles are better at consuming one resource type while adults are better at the other. The light lines represent requirements for each life stage, and the dark line represents the overall ZNGI of the species.

pact vectors of the two stages and using the typical criteria can determine conditions for coexistence between two species that have similar traits.

Ontogenetic Habitat Shifts

We've considered simple changes in size, age, and stage where organisms still generally consume the same types of resources at all life history stages even though they may vary in their relative abilities to do so. However, we can also consider organisms that undergo complete ontogenetic niche shifts (i.e., shifts in the types of limiting factors). First, we assume that there is a fixed developmental constraint so that the organism must exist in each stage for at least some period of time. We further assume that there is a limiting resource for the organism in each life stage, and so its ZNGI can be derived by combining the straight lines originating from the axes at the R^* for each stage. The combination of these lines, and thus the ZNGI for the entire life cycle, qualitatively looks like a case of a species with essential resources (fig. 5.12). With more realistic assumptions, the corner of this ZNGI is likely to be rounded as discussed previously, since success at one stage will allow the other stage to subsist on less.

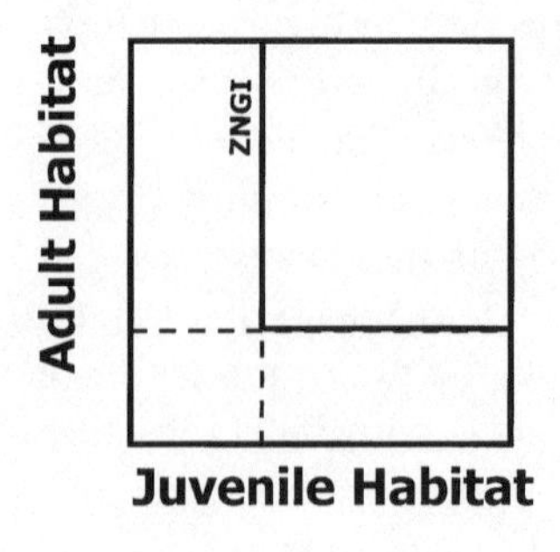

Fig. 5.12. A species with two distinct stages, adult and juvenile, each of which lives in a different habitat, and with a fixed transition time between stages. The overall ZNGI for the species (dark line) is the sum of the requirements for each stage in its habitat.

Although this prediction is not novel, there is one important point that we can derive from this model, with particular relevance to empirical research. If we were simply to measure the requirements and/or impacts of species in only one of the life stages in order to ask questions about their interactions and coexistence, we might come up with a very different answer than if we consider the entire life cycle. For example, we might measure the impacts of two interacting species within one life stage and predict that the species will not be able to coexist because one is uniformly better than another (i.e., the species do not trade off). Indeed, many empirical studies focus on only a single (or few) life stage of organisms and try to infer something about their coexistence at the population scale. In particular, many organisms have both aquatic and terrestrial life stages, but studies of species interactions, trade-offs, and coexistence often consider only one stage. However, if we simultaneously considered both life stages, we might get a very different picture.

If we assume that the developmental constraints in organisms with complex life histories are not so canalized, then a juvenile can be adaptively flexible in the amount of time it spends in one stage before transitioning to the next in relation to growth and mortality factors between the adult and juvenile habitats. Amphibians are one of the best-known groups that make these decisions as they pass through the larval (generally aquatic) to the adult (often terrestrial) stage, which often have different growth and mortality rates (Wilbur and Collins 1973; Werner 1986; Skelly and Werner 1990; Newman 1998). From a niche point of view, this argument is similar to our consideration of habitat selection behavioral decisions (without travel costs) above, but here, rather than making short-term behavioral choices, an organism makes irreversible life-stage switches (fig. 5.13a). To determine how varying the mortality and growth rates in each habitat type influences the amount of time spent in each stage, the isofitness line can be adjusted either up or down depending on the associated costs and benefits. For example, if a pond were prone to drying, or full of predators, the mortality associated with remaining there would shift the isofitness line toward the adult (terrestrial habitat) (fig. 5.13b), whereas if survival in the terrestrial habitat were chancy (e.g., due to predators or stresses), the opposite would be true (fig. 5.13c). We can then use the isofitness line to derive the ZNGI of the entire population based on each stage's ability to transition to the other habitat type. Similarly, we can use the basic criteria that we have discussed to determine the interactions and coexistence among species.

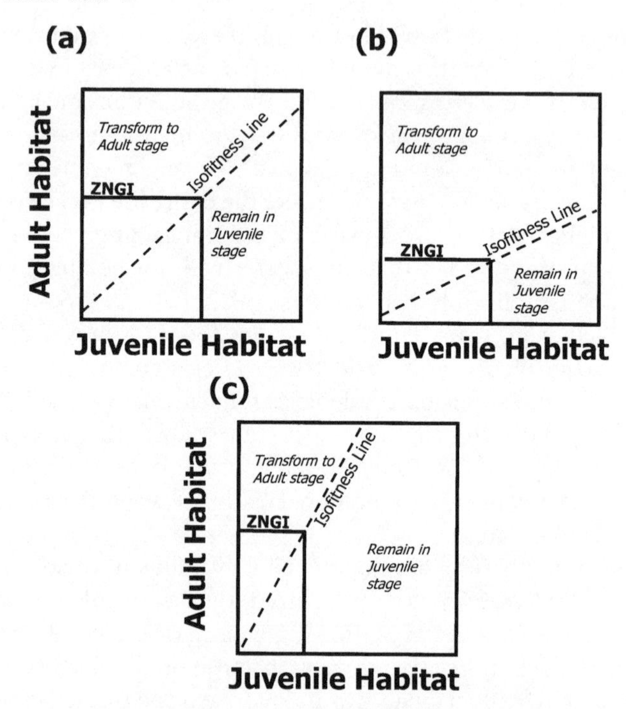

Fig. 5.13. A species with two distinct stages, adult and juvenile, where the species is plastic in its ability to transition between stages. a) The species can "decide" whether to remain in the juvenile stage or transition to the adult stage as a result of the relative costs and benefits from each habitat type. b) A shift in the cost: benefit where the juvenile environment is most costly. c) The adult environment is more costly.

Size-structured Predation

The final type of complex life history shows one way in which demographic considerations can be incorporated into the keystone predator module. We assume a simple two-stage life history, where juveniles are small and susceptible to predators, while adults are large and less vulnerable to predators. This scenario is encountered by many organisms ranging from vertebrate herbivores to gape-limited fishes.

We use the trick discussed earlier: we draw the requirement component for each life stage, where the AZNGI represents zero net fitness, while the JZNGI represents a zero net transition rate to adulthood. If juveniles are uniformly worse competitors (higher R^* for resources) than adults, the JZNGI is entirely to the right of the AZNGI (since it is also more susceptible to predators), and the JZNGI will determine whether the species can occur in that habitat. If, however, juveniles are superior competitors (in terms of the minimal amount of resources that they need to transition to the next stage) but worse at dealing with

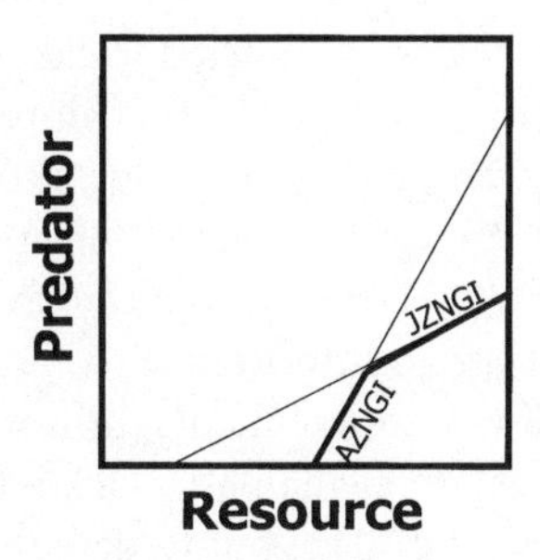

Fig. 5.14. The effects of size-structure, when the juvenile stage may be more susceptible to mortality from predators than the adult stage. Adult (AZNGI) and juvenile (JZNGI) requirements along axes of resources and predators (light lines), and the ZNGI of the species (dark line).

predators than adults, the combination of the requirements for the two stages derives the species' overall ZNGI (fig. 5.14). Thus, the species' ZNGI is constrained by the necessity for adults to be able to produce young at low resource levels and for juveniles to survive to adulthood at high predator levels. Note, however, that this simple derivation of the ZNGI can be complicated by several factors. For example, in a model with a similar basic structure, Chase (1999a) assumed that juvenile growth rates were a positive function of resource densities, which we have not done here, and derived a ZNGI with a different shape. In either case, the stability of the equilibria among species will depend on impacts; we often expect these to be such that the better defended (size-refuge) species is better predator food and the equilibrium is unstable.

5.5. Concluding Remarks

Often, ecologists focus on the complexity of natural systems. In doing so, they become so impressed with the intricacies of every species in every system that they may give the impression that nature is too inherently complex and variable to characterize according to any simple rules or generalities (see Lawton 1999, 2000, for discussions of generality in ecology). We hope to have shown in this chapter that even though there are many different types of biological realisms that characterize natural systems, many of the generalities derived from simpler models can still hold. Further, when complexities do change the nature of the interaction, they do not do so in a fundamentally unpredictable way. We suggest that rather than focusing exclusively on the complexity of every system and interaction per se, it can be productive to focus on the similarities as well.

Many of these ideas are by no means new; they are borrowed from a variety of others. We have simply tried to take these disparate frameworks and present them in a manner that illustrates the flexibility of the approach. As we have noted throughout, though, much theoretical and empirical work remains to be done in the area of adding complexities to the simple modules. For example, behavioral plasticity and resource allocation are of broad importance to ecologists, but their study has emphasized complexity rather than synthesis. Further, many types of demographic analyses are available to tackle the dynamics of structured populations (Caswell 2001), but to date there has been very little bridge between these studies and community-level processes.

5.6. Summary

1) We have shown that a relaxation of many of the simple assumptions in the previous chapters does or does not alter the general predictions. Specifically, we use case studies involving behavioral plasticity, differential resource allocation, and complex life cycles.
2) In many cases, although the shapes and positions of the ZNGIs and impact vectors change, the fundamental predictions of species interactions and coexistence remain the same.
3) In other cases, these complexities can cause nonlinearities in the ZNGIs and more complex impact vectors. This can lead to a richer array of possibilities, including the frequent prediction of multiple ZNGI intersections, leading to potential alternative stable equilibria.
4) Adding biological complexities does not limit the utility of a niche-based framework or the quest for generality. Rather than focusing on the inherent complexity of natural systems, ecologists might instead view that complexity within a broader framework.

CHAPTER SIX

ENVIRONMENTAL VARIABILITY IN TIME AND SPACE

Nature is constantly changing. On a global scale, gradual climate and geological changes continuously alter the environmental context for species interactions. On smaller scales, spatial and temporal variability changes the physical and chemical environment within which organisms live and interact.

Most of the insights that come from our approach to the niche framework occur when the dynamics are at or near a dynamic equilibrium. This does not, however, mean that the approach itself assumes equilibrium. ZNGIs represent the case where populations are at equilibrium (zero net growth), and thus the approach assumes that there is an equilibrium in the system, but this does not require that the equilibrium ever be reached. Impact vectors too, even though they are viewed near equilibrium, exist throughout the state space. Therefore, models that assume that there is an equilibrium are still relevant to systems that are in disequilibrium. However, this does mean that we need to consider a richer array of outcomes and include a different level of complexity than when this variability is not present.

The basic components of a niche-based framework, outlined in chapter 2, examined the full range of a species' birth and death rates in response to a particular environmental factor (e.g., resources, predators, stresses). However, in order to derive some basic principles of species' niches and interactions, we focused specifically on the conditions required to balance birth and death rates, from which we derive the zero net growth isoclines (ZNGIs). Of course, it is easy to conceive of two species with identical ZNGIs but with very different population growth under conditions where birth is not balanced by death; that is, in conditions with nonzero isoclines (fig. 6.1a). These two species will have different

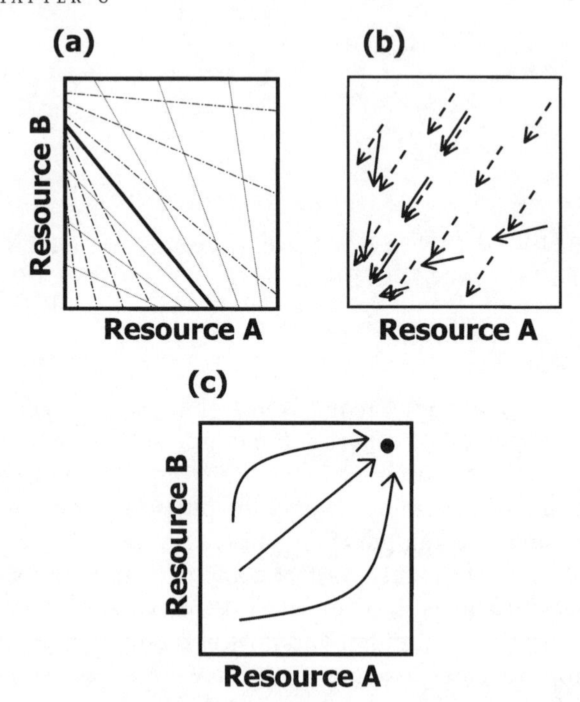

Fig. 6.1. Some examples of how variability can alter the graphical framework when species consume two resource types (A and B). (a) The ZNGIs (solid line) of two species completely overlap, while their responses (requirements) differ at other levels of fitness above and below zero (dashed and dotted lines). (b) The impacts of the two species (solid and dashed lines respectively) are identical at some points but divergent at others. (c) A field of nonlinear supply vectors.

population dynamic behaviors any time they are away from the ZNGI; thus, focusing on nonzero isoclines will be more relevant when the environment is variable. Similarly, in our previous discussions, we examined the impact component of a species' niche only at the intersection of two species' ZNGIs. However, when a consumer has nonlinear consumption rates (i.e., functional responses) along a resource gradient, the shape of the impact vector varies through the state space (fig. 6.1b). Finally, the supply vectors may diverge away from equilibrium if the resource renewal rates are nonlinear (fig. 6.1c). In the remainder of this chapter, we embrace this variability and explore the features of systems that are away from equilibrium.

6.1. Variability at What Scale?

Whether or not a system is near some sort of equilibrium will depend on what scale, both temporally and spatially, the system is viewed on (e.g., Schaffer 1981; Pascual and Levin 1999). For a spatial example, consider the invertebrate community underneath a fallen log in a forest. If we viewed the dynamics of the species under a single log, we would certainly see a highly dynamic and nonequilibrial situation as the log decays. However, we can also view the dynamics of the entire forest, where there are hundreds of similar logs. Some of these logs are new, some are old. Some have been recently disturbed, others have been left untouched for some time. If we view the average dynamics of the populations of species within the entire forest, we can begin to see a more stable situation emerging.

We can also explore how temporal averaging of a very dynamic and variable system can give rise to a situation that is reasonably approximated as equilibrial. If we view a highly ephemeral habitat such as a vernal pool for a single year, we would see that this habitat is highly variable. When these pools fill with water from rain or snowmelt, the organisms within them undergo a massive transition from dormant stages (such as resting eggs for many invertebrates or underground rhizomes for plants) to a burst of activity, growth, and reproduction. Because of the rapid generation time of many of these organisms, they may begin to stabilize and approach some sort of local equilibrium depending on the conditions of the particular year. As the summer progresses, however, and the pool begins to dry, the organisms lay new resting eggs, or otherwise go dormant, and the equilibrium seems to be lost again. However, at an even longer time scale, if we look at the same vernal pool over several years, we might begin to see a the emergence of a more regular pattern that could more reasonably be approximated as an equilibrium.

While it may be possible to view a variable system at a scale where equilibrium can be approximated, in some cases it is essential to embrace the variability of the system. For example, intrinsic variability generated by interacting organisms (e.g., population cycling or resource structuring) and extrinsic variability produced by temporal and spatial heterogeneity can both independently and interactively determine the dynamics, interactions, and coexistence of species (Chesson 2000a,b; Chase et al. 2001; Abrams and Holt 2002).

6.2. Intrinsic versus Extrinsic Variability

Many models of population dynamics show that even if an equilibrium point exists, the populations will often not settle to that point, but will instead oscillate within a stable or unstable limit cycle, or even show chaotic dynamics as a result of intrinsic variability (Rosenzweig 1971; May 1974a; Hassell 1978; Gurney and Nisbet 1998). Indeed, even some of the simplest models of population growth (e.g., logistic growth in discrete time [May 1974a]) or predator-prey interactions (Lotka 1924; Volterra 1926) can show such oscillations. Sometimes, these oscillations produce species interactions that contrast with models that predict a stable point equilibrium (e.g., Armstrong and McGehee 1976, 1980; Abrams 1998, 1999; Huisman and Weissing 1999, 2001a,b; Abrams and Holt 2002). Some types of complex attractors can even allow multiple species to coexist when they compete for a single limiting resource in an otherwise constant environment (Armstrong and McGehee 1976, 1980; Huisman and Weissing 1999; Abrams and Holt 2002). In addition to intrinsically generated temporal heterogeneity, intrinsic heterogeneity can also be generated spatially, particularly if there are trade-offs in the ways that organisms find and exploit habitats (e.g., competition-colonization trade-offs) (Levins and Culver 1971; Horn and MacArthur 1972; Wilson et al. 1999).

Extrinsic variability is imposed by the environment and has often been used to explain why many species can coexist on a few resources (Hutchinson 1961; Levin and Paine 1974; Huston 1979; Chesson and Huntly 1997; Litchman and Klausmeier 2001; Chesson 2000a,b). The issue was catalyzed by Hutchinson's (1961) "paradox of the plankton" to explain the coexistence of many species on a few resources as the result of temporal fluctuations that occurred on time scales faster than those for competitive exclusion. However, it is now becoming well understood theoretically that environmental fluctuations and disturbances do not necessarily slow the rate of competitive exclusion (Chesson 1991, 2000b; Goldwasser et al. 1994; Chesson and Huntly 1997; Wootton 1998). Instead, the prevention of exclusion occurs because some species benefit disproportionately from these fluctuations relative to others. Species coexist due to trade-offs between their relative competitive abilities and their abilities to deal with these fluctuations (Chesson 1991, 2000b; Chesson and Huntly 1997; Wootton 1998). The same qualitative generality is true for variability in space, where species trade off in their ability to compete for resources within a patch and to find new patches (Yu and Wilson 2001).

There are also cases where intrinsically and extrinsically generated variability will interact (Tilman 1994; Huisman and Weissing 2001a,b; Chesson 2000b; Chase et al. 2001; Levine and Rees 2002). This interaction can either enhance coexistence by generating new opportunities for species to coexist if they have trade-offs for dealing with the different types of heterogeneity, or it can inhibit coexistence if the two types of heterogeneity negate the coexistence that was possible with one or the other type of variability.

We will modify the use of the niche to address the role of these trade-offs. There are several methods that allow us to investigate species' interactions and the potential for coexistence within variable environments. However, our inability to fully address population dynamics during the transient phases associated with fluctuations will limit our conclusions somewhat.

6.3. Species Interactions and Coexistence in Variable Environments

To orient our thinking we start with a brief overview of an analogous approach developed by Chesson (Chesson 1994, 2000b; Chesson and Huntly 1997). Chesson used an analytical model based on the ability of populations to grow when rare, which he termed $\bar{r}(i)$. In our graphical approach, this is similar to the environmental conditions for the requirement component of a species' niche, where the species' births exceed deaths. Three components characterize $\bar{r}(i)$: (1) $r(i)'$, which describes the deterministic responses of species in the absence of environmental fluctuations, (2) ΔN, which represents intrinsically generated variability (nonlinearities due to interspecific interactions) and (3) ΔI, which represents the interaction between effects of extrinsically generated temporal variability and the effects associated with the absence of those temporal fluctuations (sometimes called the "storage effect"; Chesson and Warner 1981; Warner and Chesson 1985).

To describe the regulation of a population $\bar{r}(i)$, Chesson (2000b) used an equation with these three components:

$$\bar{r}(i) = r(i)' - \Delta N + \Delta I.$$

Thus, this equation shows how the outcome of interactions in a nonequilibrium situation (including ΔN and ΔI) can deviate from their outcome in the equilibrial model of that situation (only including $\bar{r}(i)$), since the second two terms are ignored in the equilibrial model. In order to understand the outcome of the model, it is necessary to understand the processes that regulate the equilibrium point. Chesson (2000b) points out that the mechanisms that regulate coexistence in the absence of fluctuations, do not go away, or necessarily change, in the presence of fluctuations (see also Chesson and Huntly 1997). As a result, the argument that fluctuations prevent competitive exclusion, and thus enhance species coexistence, is inadequate unless these fluctuations also alter ΔN or ΔI. That is, the species must differentially respond in these components to the presence of the fluctuations and have traits that allow them to deal with the fluctuating environment differently. This basic insight is also at the crux of the distinction between the effects of environmental variability on niche and neutral theories of species interactions. While Chesson's niche-based model emphasizes that differences among species are needed in order for diversity to be maintained, Hubbell's (Hubbell 1979, 2001; Hubbell and Foster 1986) neutral model assumes that species are equivalent (i.e., do not trade off) and predicts that regardless of the degree of environmental variation, species will always be headed toward extinction. In the neutral model, then, diversity can be maintained for long time periods only as long as the speciation rate, or immigration rate from some external source, is at least as high as the rate of extinction.

Chesson's (2000b) equation draws our attention to the fact that fluctuations can promote coexistence if (1) species have nonlinearities in their competitive effects that cause intrinsic variability (ΔN) and different species are favored at different points along the fluctuation, or (2) there is extrinsic variability (ΔI) and species differ in which points along the fluctuation they find favorable. Chesson (2000a) has also shown how the concepts of temporal variation can be extended in a similar (although different mathematical) way to the concepts of intrinsic and extrinsic heterogeneity in space (see also Hastings 1980; Schmida and Ellner 1984; Pacala and Tilman 1994; Tilman 1994; Chase et al. 2001; Mouquet and Loreau 2002; Levine and Rees 2002).

A full accounting of spatial and temporal variability is more complex in dimensionality and in parameterization than we can accomplish using the graphical methods advocated in this book. For example, the ΔN and ΔI terms in the models are somewhat vague and phenomenological, which makes them general but not very useful for any given situation. For this reason, it is possible to derive several more specific mechanistic equations for the ΔN and ΔI terms for a variety situations (e.g., competition involving annual plants, perennial plants, seed banks). These are generally complex and can be highly nonlinear. Similarly, complex nonlinear models also generate highly parameter-dependent outcomes in many multispecies models (Abrams 1998, 1999; Huisman and Weissing 1999, 2001a,b; Abrams and Holt 2002). Our goal, however, is to incorporate Chesson's basic insight—that some of the major consequences of nonequilibrium situations depend on the presence of relevant trade-offs among interacting species—into the graphical niche framework that we emphasize. This framework offers simple ways to express these trade-offs and make predictions about the consequences of environmental fluctuations and spatial heterogeneity, which will allow more general insights into both theory and nature.

Environmental Variability as Distributions of Supply Points

Perhaps the simplest way to evaluate variable environments is to imagine that what happens under different environmental conditions, either spatially or temporally, is approximately the sum of what happens under each of the individual conditions. Tilman (Tilman 1982, 1988; Tilman and Pacala 1993) represented spatiotemporal variation in resource supplies as a set or range of resource supply, rather than a single supply point (fig. 6.2a) (see also Levin 1974). That is, in a heterogeneous environment, if there is some place or time where a particular supply exists, then the outcomes of species' interactions predicted at that supply may occur. This cloud of supply points can represent either spatial or temporal variation in resource supply, and a species is often expected to exist (or coexist) when this cloud overlaps the ranges where it is predicted to do so.

We can make this trick slightly more precise. For spatial variability, this approximation is likely to be most appropriate when (1) all individuals have access to the entire space, (2) organisms find their optimal habitat without a significant foraging cost, and (3) species that are more numerous, due to greater availability of the habitat they find favorable, do not drive less numerous species in rarer habitats to extinction. In these conditions, the outcome of species' interactions in multiple patches is the sum of all the outcomes that would occur in each of the individual patches. For example, if five patch types are equally frequent with supply points as shown in fig. 6.2b, the models predict that only pairs of species will occur in each patch, but that the larger community will have all six species present.

For the temporal case, species must have mechanisms that allow them to persist during times when they are inferior competitors (Chesson 1985). Mechanisms that would allow for such a storage effect (Chesson and Warner 1981; Warner and Chesson 1985) include seeds that can be dormant (e.g., Pake and Venable 1996), resting eggs of several aquatic invertebrates (e.g., Caceres 1997), invulnerable and long-lived life history stages, or other dormancy mechanisms (e.g., hibernation). We can illustrate this by a simple example (fig. 6.2c) where two species compete for two resources in a variable environment. Temporal variability causes the supply point to be shifted through time. At different points in time, different outcomes are expected, but in the long term, we would predict that the two species

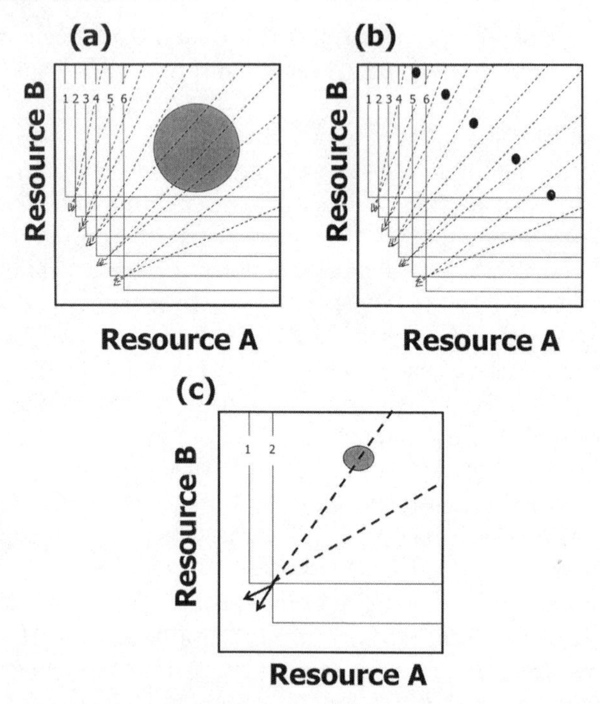

Fig. 6.2. Some tricks to incorporate spatiotemporal variability in environmental resource supply. (a) Six species competing for two essential resources. The shaded cloud represents all of the environmental supply of resources that are present at some time or place. (b) Variation in the resource supply at five localities within a region. The species that can exist in the region will be all that can live in any locality. (c) The shaded cloud represents temporal variation in resource supply and partially overlaps the zones where the species coexist and where one species outcompetes the other.

should coexist in this environment since the set of supply points includes both coexistence and exclusion. However, this does not have to be true. If the species do not have a storage effect and the population dynamics occur on a fast time scale (relative to the time scale of environmental change), then we would predict that one species would drive the other toward extinction when the supply point moves out of the zone that predicts coexistence. Even though the excluded species might be able to recolonize when the environment shifts to more favorable conditions, this would presumably take some time unless there is a nearby source population.

This approach can serve as a useful heuristic guide for empirical research and for hypothesis development (e.g., Tilman and Pacala 1993; Leibold 1996), even if it is sensitive to criticism (see Abrams 1995, 1999). For more detailed information on quantifying how variability influences species coexistence, we refer the reader to Chesson's work (Chesson 1994, 2000a,b; Chesson and Huntly 1997).

Treating Variability (and Higher Moments) as Niche Axes

Another approach is to decompose the effects of a particular environmental factor into a mean effect and a variance effect (Levins 1979). We develop this situation here by considering competition for a single resource, but similar reasoning could be used for other cases, such as predation. Here we use one niche axis (the horizontal one) to evaluate how each species responds to differences in mean resource levels, holding the variance in resource levels constant. And we use the other axis (the vertical one) to evaluate how each species responds to differences in variance in resource levels, holding the mean resource level constant. Obviously, species could differ in both of these responses in a variety of ways.

We illustrate a case where there are trade-offs associated with a species' ability to deal with variance and to compete for resources in the absence of variance. By analogy with behavioral ecology theory, we discuss three types of species that differ in their responses to resource variance: risk-averse, risk-neutral, and risk-prone (Caraco 1980; Real and Caraco 1986). The risk-averse species (species 1) has the lowest resource requirements in the absence of variability, but is more negatively affected by variation in resource levels (fig. 6.3). For example, this species could have a limited ability to store nutrients, and would then be negatively affected by temporal variation that leads to occasional lower resource levels. The risk-neutral species (species 2) is unaffected by environmental variability but has higher resource requirements in the absence of variability. For example, this species could have the ability to store nutrients, but would then also have the cost associated with this trait. The risk-prone species (species 3) benefits from variable resource availability, and has a ZNGI with a negative slope. For example, this species could have a large capacity to store excess consumption obtained during periods of high resource availability for use during periods of low resource availability.

Consumers in such a system impact both the mean and variability of the resource levels. For example, a consumer can decrease the variance in resource levels by rapidly consuming pulses in resource availability or by facilitating regrowth of resources when they get very low (via recycling). Conversely, a consumer can enhance the variability in resource levels. For example, many predators can increase the amplitudes of fluctuations of their prey species.

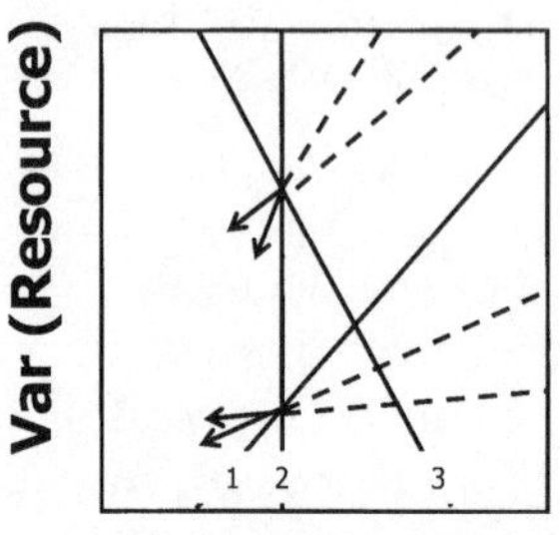

Fig. 6.3. Three species responding to the mean resource abundance (*x*-axis) and the variance in the abundance of that resource (*y*-axis). Species 1 responds negatively to the variance and is risk-averse, species 2 is unaffected and is risk-neutral, and species 3 is advantaged in variable environments and is risk-prone. Each species also can impact the resource variance. The ZNGI intersections between 1 and 2 and 2 and 3 are uninvasible, while the equilibrium between 1 and 3 is invasible by 2, and we do not consider the impacts at this equilibrium. Criteria for coexistence and exclusion are predicted in the typical way.

As in our earlier models, two linked trade-offs are required for coexistence in a variable environment. First, we would expect that species with the lowest resource requirements in the absence of variance (lowest R*) would have to be sufficiently negatively affected by variance relative to their competitor (i.e., have a shallower ZNGI), such that the ZNGIs of the two species would cross. Second, we would expect that local coexistence would be possible only if the species that most benefits (or is least harmed) by variance has a greater effect in reducing the variance in resource levels than its competitor. If the opposite of this latter condition is true, than the species will not coexist locally, but alternative stable equilibria can allow them to coexist regionally.

We also predict that the outcome of interactions between species would depend on both the supply of resources and the environment's supply of variability in the absence of these consumers. In environments that have low variability, we would expect that the more risk-averse species, with lower resource requirements, to exclude the more risk-prone species. In environments with very high variability we would expect that the more risk-prone species would exclude the more risk-averse one. Finally, we expect coexistence when variability is intermediate, if the more risk-prone species reduces the variance more than the risk-averse one, or we expect an unstable equilibrium if the converse is true.

Interestingly, these results are similar to those obtained by Chesson using both a generalized approach to temporal variability (Chesson 2000b) and more specific models (Chesson and Warner 1981; Chesson 1985, 1994). Our graphical approach complements Chesson's and illustrates some common dependencies on linked trade-offs for coexistence. Further our graphical approach may provide a more transparent link to empirical studies.

6.4. Concluding Remarks

In this chapter, we have discussed some ways to apply methods to explore how variability, through both time and space, will influence the predictions of species' interactions within the niche framework. Fully accounting for the effects of particular types of fluctuations will probably require more rigorous theoretical models than we can incorporate into this approach, but at least qualitatively, our methods can provide predictions that are similar to the more rigorous approaches. We believe that these methods will provide a useful peek into the complexities of variable and nonequilibrial systems. Furthermore, these models may be more useful to empiricists than the array of complex analytical and simulation models due to their flexibility and accessibility in developing hypotheses and making broad comparisons.

6.5. Summary

1) The graphical niche framework does not assume equilibria are always achieved, but rather assumes there is an equilibrium present.
2) In some cases, spatial and temporal variability in environmental characteristics can complicate the simple equilibrial predictions.
3) Processes that might not appear equilibrial at one time or spatial scale might take on many more equilibrial attributes at other scales.
4) Intrinsic and extrinsic fluctuations can affect species' interactions separately and interactively. The fluctuations may provide avenues for trade-offs among these species, potentially allowing them to coexist.
5) Variability can be incorporated into the niche framework in two different ways: first, by viewing the environmental supply as a range of points that incorporates all resource levels organisms experience in space or time, and second, by treating variation explicitly as a niche axis where a species responds to both the mean and the variability of a particular factor.

CHAPTER SEVEN
SPECIES SORTING IN COMMUNITIES

The niche concept has its origins in community ecology. Grinnell, Elton, Gause, Hutchinson, MacArthur, and others who promoted the niche concept were mainly interested in how the niche of a species influenced its interactions within communities (see reviews in Whittaker et al. 1973; Hutchinson 1978; Giller 1984; Schoener 1989). As we pointed out in chapter 1, some of the most important topics in community ecology in the 1960s and 1970s, such as competition and coexistence, were inextricably linked to the niche concept. Here, we expand our view of the niche beyond the realm of a few species and their interactions and discuss ways in which the basic niche framework discussed thus far can be extrapolated to more realistic complex communities. Ecologists have wrestled with a number of methods to study these more complex communities, each of which has its own strengths and weaknesses. Our goal here is to illustrate how the niche concept, as we have developed it so far, can contribute.

In many ways, most of the work we have laid out in the previous chapters has been directed at developing a framework with enough flexibility to deal with questions related to more complex communities while retaining insights and knowledge about the natural history of organisms. In this chapter and the next one we extend simple modules to generate approximations of more complex situations. In doing so, we consider how larger-scale (or emergent) phenomena about communities might be related to species interactions.

7.1. The Problem of Complexity

To this point, we have primarily focused upon a very limited set of interactions in both theory and empirical examples. Indeed, even the most complex theoretical models of species interactions include only a small fraction of the total number of species in a community, and empiricists usually limit their focus to a group of interacting species within a much more complicated community.

The open water of lakes and ponds has been heralded as a system in which the entire community can be explained fairly well by simple models (see Strong 1992). Models of resource competition (Tilman 1982; Huisman and Weissing 1999), food chains (Carpenter and Kitchell 1993), and keystone predation (Leibold 1999) have all been used to explain aspects of community structure within these biotas. The success of these relatively simple conceptual frameworks has caused some to claim that these communities are inherently simpler than their terrestrial analogs, and even simpler than other aquatic communities (Strong 1992; Polis and Strong 1996; Polis 1999; Polis et al. 2000). In reality, both lakes and ponds can contain tens to hundreds of plant, herbivore, and predator species, many of which are difficult to discern because of their small size, but they nevertheless constitute complexity as perceived by the food web (reviewed in Chase 2000). Perhaps then the successes of the simple frameworks at predicting important patterns in aquatic systems result because often aquatic ecologists are more likely to lump species into categories, whereas terrestrial ecologists have been more reticent to do so (but see Oksanen and Oksanen 2000).

Even experimental microcosm communities are probably more complex than many researchers would like. For example, studies of protists have served as a staple for evaluating and refining community ecological theory for a long time (e.g., Gause 1936; Drake et al. 1996; Lawler 1998). However, researchers are rarely able to control the diversity of the microbial communities on which the protist food web is based. Studies examining bacteria as a model system are also more complex than the simple modules because bacteria can rapidly evolve traits and diversify (Bohannan and Lenski 1997; Rainey and Travisano 1998; Buckling et al. 2000).

In chapter 4 we highlighted some empirical success stories, which involved studies of interactions among a relatively small numbers of species. We believe that these studies were successful largely because the researchers were able to concentrate on dominant species in their ecosystems and a single community module (e.g., seed-eating rodents, plants that compete for a few limiting resources). Nevertheless, we know from theory (e.g., Schaffer 1981; McCann et al. 1998; McCann 2000) and from experiments (e.g., Leibold and Wilbur 1992; Lawler and Morin 1993; Wootton 1994) that the addition of a single species can potentially alter the pattern of species interactions in the rest of the community. The question is how to incorporate the knowledge of natural history that is implied in a niche-based approach to generate hypotheses and elucidate larger-scale patterns that result at the level of the entire community.

We have stressed an explicit and mechanistic approach to species interactions in previous chapters. How can we apply this approach when we can't even list all of the interactions among species within a community, much less pretend to be able to quantify the shapes of their ZNGIs and impact vectors? One way that ecologists have tried to tame the complexity associated with highly diverse natural communities is by lumping taxonomically or functionally similar species, particularly based on their feeding relationships, into one of various groupings. These categories include: (1) *trophic levels,* where organisms are categorized according to whether they primarily consume energy from the sun (i.e., plants), from plants (i.e., herbivores), or from animals (i.e., carnivores), (2) *functional groups,* where within trophic levels, organisms perform functions in similar ways (e.g., N-fixing plants, benthic herbivores), and (3) *guilds,* a concept similar to functional groups, but often also associated with a taxonomic grouping (e.g., seed-eating rodents, phloem-feeding insects). Similarly, in food web studies, the term "trophospecies" has been used to define a group of species that have identical qualitative linkage patterns in a food web (Yodzis 1988, 1993). Because the niche requirements of species within a community probably occur along a continuum rather than in discrete categories, these concepts are often imprecise and somewhat subjective. Nevertheless, when communities are composed of tens to hundreds of species, discrete categories can allow insight into what might otherwise be seen as incomprehensibly complex.

Community ecologists ask diverse questions (see, e.g., Morin 1999). It would be beyond the scope of our book to even try to comprehensively discuss how the niche framework relates to all aspects of community ecology. Instead we illustrate how the niche concept can be used to address some of these questions. These will include an examination of species distribution along environmental gradients, the similarity of coexisting species (this chapter), the role of species in community succession and assembly, and the patterns of biodiversity (chapter 8). First, we discuss the process that we refer to as *species sorting,* which lays the foundation for several questions of community and ecosystems ecology.

7.2. Species Sorting within Groups of Functionally Similar Species

Species sorting is a process by which locally coexisting species, which are a subset of the potentially larger regional pool, come to exist in a particular locality. To illustrate this, consider what happens when we have a diverse group of N species that share niche axes. We base this discussion on three community modules, but note that other modules will have qualitatively similar predictions: (1) resource competition, where species compete for two shared essential resources, (2) keystone predation, where species compete for shared resources and shared predators, and (3) resource-stress interactions, where two species share a resource and are limited by an environmental stressor.

We assume for simplicity that the equations for all N species have the same form (e.g., linearity) so that the general shape of the ZNGIs is similar. We also assume that local population dynamics equilibrate relatively quickly, so that there are not more than two species that can coexist locally (since there are only two limiting factors). We discuss deviations from this assumption below. We consider how an array of six species might interact (fig. 7.1). Species 1 through 5, whose ZNGIs appear as solid lines, exhibit perfect trade-offs—each species is slightly better at dealing with one factor and slightly worse at dealing with the other. Species 6, whose ZNGI is a dashed line, does not follow this perfect trade-off and is always inferior to at least one other species in its ability to deal with each factor.

Ignoring the relative impact vectors for now, we can already discern several aspects of coexistence and invasibility among pairwise combinations of species. In each case, an equilibrium occurs where the ZNGIs of any two species intersect, and the stability of those equilibria will be determined by the relative position of the impact vectors. However, not all of these equilibria are equivalent. The relative position of the potential equilibria points among pairs of species determines whether they are invasible. Uninvasible equilibria (denoted by filled circles in fig. 7.1) are those in which the limiting factor at this intersection is maintained at a lower level (in the case of resources) or at a higher level (in the case of predators or stresses). Here, no other species in the pool could exist at lower levels of resources or higher levels of predators and/or stresses. Conversely, equilibria that can be invaded by at least one other species are those that occur interior (in the case of resources) or exterior (in the case of predators/stress) to these points. Such invasions would lead to extinctions by at least one of the original species.

For example, imagine a local community that consists only of species 1 and 5. Now, if we allow all species from the regional species pool access to this local community, it is easy to see that the 1-5 species pair can be invaded by any of the species 2, 3, or 4. The invader will drive either 1 or 5 extinct, depending on the resource supply. This is

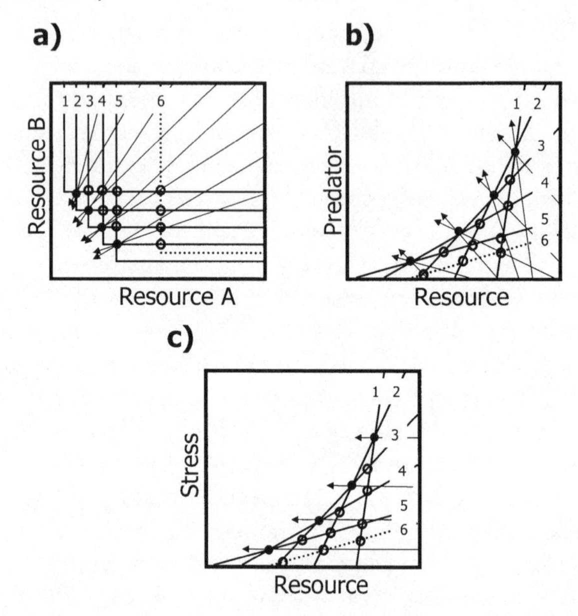

Fig. 7.1. Examples of the regional species sorting process. (a) Competition for two essential resources. (b) Competition for a resource with a common predator. (c) Competition for a resource with a shared stress. The ZNGIs for species 1–5 show perfect trade-offs along the two axes, while species 6 is a relative "dud." Filled circles represent uninvasible equilibria, while open circles represent those that are invasible by at least one other species. Impact vectors and their inverses indicate zones of resource supply where species pairs potentially coexist as before.

what we mean by the idea of *species sorting*. We can follow this idea through even more to consider all of the other pairwise combinations. The result is that in fig. 7.1 there are only four possible uninvasible pairwise combinations of species that can occur (solid circles): 1-2, 2-3, 3-4, and 4-5. Each of the ten other possible pairwise combinations (open circles) of species can be invaded by at least one species.

Five of the six species in fig. 7.1 are associated with at least one of the uninvasible combinations (species 1, 2, 3, 4, and 5), but one species, 6, is not associated with any of them. We refer to this species as a "dud" because we predict it will go extinct from the regional species pool. One characteristic of this species that makes it a dud is that its ZNGI does not cross the ZNGIs of *all* the other species. Thus, it will always be displaced by another species and cannot persist in any local community, regardless of environmental supply. Alternatively, if a species were superior to all other species at utilizing both factors, which is not shown in the figure, this species would eliminate all other species from the regional species pool. In this case, all the other species would be the duds.

So far, species sorting has had two effects. First, it has culled out all the duds from the regional biota. This results in a regional species pool that is self-selected to show a strong trade-off in its members' relative responses to both factors. In addition to our discussion of this trade-off at the local scale in chapter 2, we now predict that this trade-off also applies at the regional scale. Second, it has constrained the set of locally coexisting species pairs from fourteen potential equilibria to only four uninvasible ones. In addition, since species will not always coexist stably, each of the five species that aren't duds can also exist alone under certain resource supply conditions. Combined, this leads to the prediction that the process of species sorting, as we describe it here, reduces the number of communities consisting of one or two species that can occur from nineteen to only nine.

In addition to being influenced by the relative positions of species' ZNGIs, species sorting is also affected by processes related to species impacts, just as it is when only considering local coexistence. Specifically, these impacts further constrain the range in which species can coexist, except in the case of stress (fig. 7.1c) because species have no impact on the level of stress. The impact vectors of the species influence the consequences of species sorting in three ways. First, a given species combination is stable and uninvasible under a limited range of supply. Second, different supply conditions will produce different sets of species combinations in different local communities. Third, these species combinations can involve either stable coexistence of pairs of species or unstable equilibria, depending on whether impacts are positively or negatively correlated with requirements.

Species sorting could have profound consequences for community-level phenomena and is often ignored in other approaches. In addition, there are some consequences of species sorting that we do not consider further here. For example, Abrams (1998) presented a theoretical analysis showing that if competing predators can drive one species of prey extinct, reducing the diversity of resources, this can profoundly influence their competitive interactions.

7.3. Effects of Species Sorting on Community Phenomena

In the remainder of this chapter and in chapter 8, we use the process of species sorting and the niche-based framework to address several issues of community ecology. In this chapter we ask (1) how interacting species are distributed along environmental gradients, (2) whether species exist in nested subsets, or as gradient replacements along environmental gradients, and(3) how niche relations determine patterns of ecological similarity among coexisting species. In the next chapter, we ask (4) what roles species interactions play in community succession, (5) how communities assemble, and (6) how species' interactions regulate patterns in the distribution of biodiversity.

How Are Interacting Species Distributed along Environmental Gradients?

Although we did not explicitly discuss ecological gradients when we introduced the graphical niche framework, the connection is obvious. The axes of factors along which species interact, be they resources, predators, or stresses, can be used to envision a gradient in these factors when we expand our view to a regional spatial scale. If the regional pool of species consists of species with traits that are sorted along some environmental gradient, then the environmental conditions within any locality will favor different sets of species. That is, along an environmental gradient, we are essentially exploring how changing resource (or predator or stress) supply would alter the species composition along that gradient. Below, we briefly discuss three examples of how this can occur.

First, in the simple two-resource competition model, the total availability and the ratio of the two different resources in a given locality will determine which species or groups of species are expected to exist at that locality (Tilman 1982, 1988; Tilman and Pacala 1993). Thus, if we were to walk among localities that varied in the relative ratios of different soil resources, such as phosphorus relative to nitrogen, we would expect a predictable change in the relative composition of the plant species in those localities. The idea that plant species distributions along environmental gradients can be predicted by the depiction of their relative requirements for limiting nutrients (or light) has received considerable empirical attention, especially in phytoplankton (Tilman 1977; Smith 1983; Sommer 1990; Watson et al. 1992), old-field plants (Goldberg and Miller 1990; Wedin and Tilman 1993; Inouye and Tilman 1995), and emergent wetland plants (Grace and Wetzel 1981; Weiher and Keddy 1995).

Second, species with a common predator might be expected to distribute themselves along a productivity gradient according to their abilities to consume resources relative to their susceptibility to predators (Leibold 1996). At relatively low levels of resource supply, species that are better resource competitors (at the cost of defense against their enemies) are expected to dominate. Alternatively, at high levels of resource supply, species that are better defended from predators (at the cost of relative competitive ability) are expected to dominate. These predictions are supported in regional surveys of freshwater ponds where species that were less susceptible to predators increased in frequency as the productivity of those ponds increased (Leibold 1999).

Third, species can distribute themselves along gradients of environmental stress when they also compete for some other limiting factor, such as resources. Wellborn et al. (1996) reviewed freshwater pond permanence: some ponds dry yearly (e.g., temporary ponds, vernal pools), while others rarely dry. They suggest that a common pattern is that many species of animals (invertebrates and amphibians) living in temporary ponds tend to be worse resource competitors and less defended against predators, while species that live in more permanent ponds were worse at dealing with pond drying. Likewise, Tilman (1994) showed that plant species that were better able to colonize fields following disturbance were poorer resource competitors.

We return to these predictions in chapter 8, when we discuss issues of species diversity along these gradients. For now, we can draw three main predictions of how species sorting influences the distribution of species along environmental gradients. First, species should distribute themselves along an environmental gradient in direct relation to their relative requirements for the factors of interest. That is, species whose relative requirements for a particular factor are low should exclude other species in environments where the relative supply of that factor is low, and so on. Second, species should exist in fewer sites than they could potentially occupy as a result of competitive exclusion by other species. Third, along simple environmental gradients where species utilize two factors, the range of environmental conditions that one species uses should not be nested within that of another species, or the species with the narrower range would be competitively excluded. If such nesting occurs it would most likely be due to some unmeasured covarying factor.

If the relative impacts of species are greater on the factor that least limits them, then species are not predicted to coexist locally, and alternative monocultures of a species will exist. Under these circumstances, we would make the same general predictions about species distributions along a regional environmental gradient. However, locally we would expect a considerable amount of variation; different species would dominate different localities even with the same resource supply.

Do Species Exist in Nested Subsets, or as Gradient Replacements along Environmental Gradients?

A question closely related to how species are distributed along environmental gradients dates back at least to the time of Whittaker's (1956, 1962, 1967) pioneering studies of how plant species were dis-

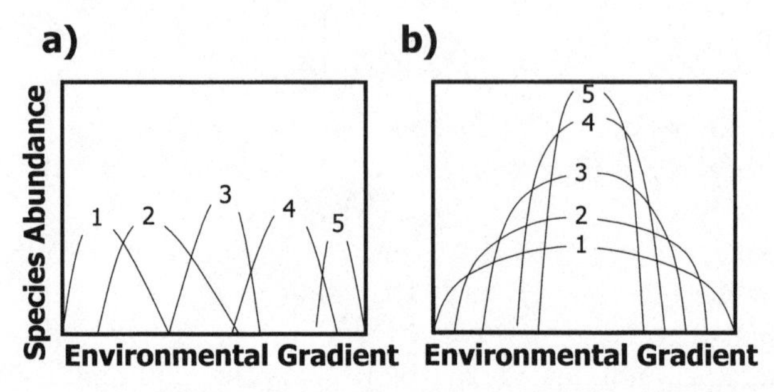

Fig. 7.2. Two hypothetical distributions of five species along an environmental gradient. (a) A gradient replacement, where one species replaces the next along a gradient. (b) A nested subset, where the species with the narrowest distribution is nested within the distribution of the next most distributed species.

tributed along elevational gradients in the Smoky Mountains of Tennessee. Whittaker was interested in whether species were replaced by other species (a pattern referred to as a *gradient replacement*) (fig. 7.2a), or whether new species were added but few (or none) were lost (*nested subsets*) (fig. 7.2b).

So which pattern, gradient replacements or nested subsets, is observed in nature? Well, as with most questions in ecology, it depends. This is why there has been much discussion about the existence of, and statistical tests for detecting, nestedness (e.g., Patterson 1987; Worthen 1996; Wright et al. 1998; Jonsson 2001; Leibold and Mikkelson 2002). Surprisingly, however, little attention has been paid to the specific mechanisms that might predict the occurrence of nestedness or gradient replacements. More optimistically, the process of species sorting gives us a way to predict the conditions under which we might expect each pattern to manifest itself.

For the case of resource competition, we consider qualitatively different scenarios. First, in fig. 7.3a, we assume that resource ratios are near 1:1. When the regional variance in resource supply (denoted by a clouded distribution to represent variance in environmental supply, as in chapter 6) does not change as the average supply of resources increases (dark circles), we expect no shifts in species composition or diversity with resource supply. On the other hand, if the regional variance in resource supply increases with increasing average resource supply, we expect the number of species that can coexist at equilibrium to increase with the mean resource supply. This is because the resource supply cloud occurs in the range of existence for more species at higher resource levels. Along this resource gradient, species are added, but no species are lost, which is consistent with a nested subset.

Alternatively, if we assume that the resource-ratio is constant, but

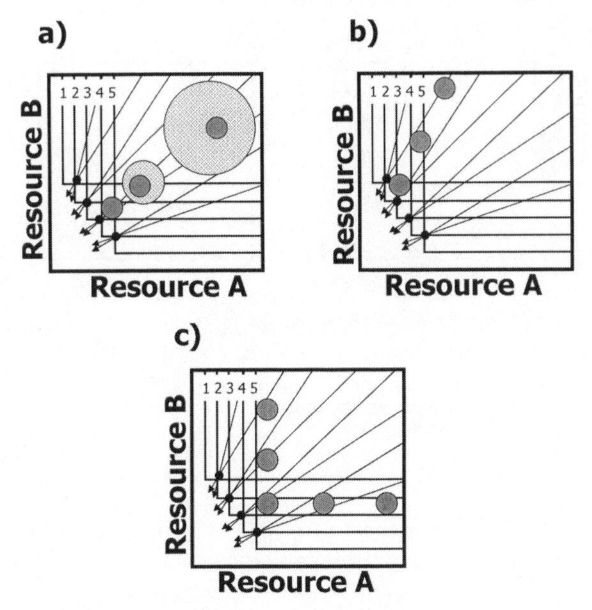

Fig. 7.3. The case where species (five in this case) compete for essential resources and the supply of resources vary. ZNGIs and impacts are the same as those for species 1 through 5 in fig. 7.1a. (a) Resource ratios are near 1:1 and do not vary as the absolute level of resources vary. The darker, smaller cloud of supply points represents the case where the heterogeneity in resource levels does not vary with the absolute level of resources, while the lighter clouds that increase with size with increasing levels of resources represent the case where heterogeneity increases with magnitude. (b) When resource ratios are strongly skewed from 1:1 but remain constant with increasing resource supply. (c) When resource ratios vary; the level of supply of one remains the same while the other changes.

considerably different from 1:1 (fig. 7.3b), or the resource ratio varies such that the supply of one resource increases while the other remains the same (fig. 7.3c), we predict a different outcome. In both these cases, we predict that different species will be present at different resource supplies, regardless of the level of heterogeneity, which is consistent with gradient replacement.

In a similar manner, we can predict what would happen to species composition along environmental gradients when species share resources and predators (fig. 7.4a). In this case, we do not need to consider the ratio of the supply of resources and predators, since we assume that predators cannot vary independently of prey availability. Regardless of whether heterogeneity is constant or varies, we predict that species composition will change along a resource supply gradient. Specifically, at low resource supply, we predict that only those species that can exist on depauperate resource conditions will persist. How-

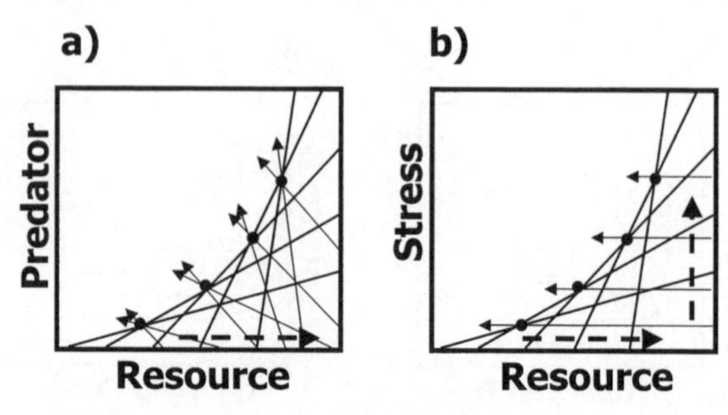

Fig. 7.4. Species composition changes along gradients. (a) Five species along predator-resource axes (same as in fig. 7.1b), where the level of resource increases along the *x*-axis only (depicted by the dashed horizontal arrow). (b) Five species along stress-resource axes (same as in fig. 7.1c), where the level of resource changes horizontally (horizontal dashed arrow) and the level of stress varies vertically (vertical dashed arrow).

ever, as we increase the resource supply, we also increase the intensity of predation, due to the interdependence of the food web, and we expect species that are less susceptible to predators to be favored. This is a gradient replacement.

Finally, we consider the case where species are distributed along a gradient of resources and stress (fig. 7.4b). In this case, both resources and stress can vary independently from one another. If we vary resources in this scenario (along the x-axis) (horizontal arrow), we do not expect any shifts in species composition, regardless of the magnitude of heterogeneity or the magnitude of the stress. This is because at a particular level of stress, there will be a single species that will be the superior resource competitor regardless of the resource supply. However, if we vary the magnitude of the stress (vertical arrow), we again see a gradient replacement.

So, we can quickly see that the specific assumptions one makes, including the limiting factors, the requirements and impacts on those factors, and the environmental supply, will lead to a rich array of predicted patterns. But how does this relate to what we see in the real world? It is perhaps too early to use empirical data to evaluate these hypotheses. Whittaker's studies along an elevational gradient could be considered either a productivity or a stress gradient or some combination of both. In either case, we would often predict that he would find a gradient replacement, which he did. However, many studies, particularly on islands, have found nested subsets of species distributions (e.g., Lomolino 1996). These patterns are likely to be influenced not only by local conditions of the islands, but also by patterns of dispersal limitation and other biogeographic processes. Thus, the theories we have discussed above are inadequate to fully explore this situation. In chapter 11, we will initiate a discussion on other ways that regional processes can be incorporated into models of local community interactions (see also Loreau and Mouquet 1999; Amarasekare and Nisbet 2001; Mouquet and Loreau 2002) in the context of metacommunities.

How Do Niche Relations Determine Patterns of Ecological Similarity among Species?

Gause's axiom that no two species can have the same niche and coexist has motivated a lot of predictions about the distribution of species. But following Gause, and especially motivated by Hutchinson's (1959) discussion of the limits to species similarity and therefore diversity, ecologists began to wonder just how similar species could be and still coexist. The idea that the traits of species coexisting locally should have a limit to their degree of similarity, sometimes referred to as *character displacement,* was central in the development of community ecology through the 1960s and 1970s and was an important component of what we previously described in chapter 3 as conventional niche theory. Even today, prominent ecologists strongly disagree as to whether ecological (niche) differences among species are necessary in order for species to coexist locally (see Chesson and Huntly 1997 and Chesson 2000b versus Hubbell 1997, 2001).

There are two distinct meanings to the term "character displacement." The first, based on precedence, is evolutionary and involves the idea that when two similar competing species coexist, they will evolve to diverge in some important characters or traits, at least up to a point (see Brown and Wilson 1956 for the term's original usage). We defer most discussion of this issue until chapter 10. The second meaning of "character displacement" predicts that any pair of coexisting species will be more different from one another than any pair of species chosen at random from a regional species pool (Leibold 1998). We refer to this idea as *community-wide character displacement* (CWCD).

CWCD is controversial in community ecology and was at the core of the antithetical attacks on the niche concept described in chapter 1 (Gotelli and Graves 1996). Following the ideas of Lack (1940, 1947), Huxley (1942), and especially Hutchinson (1959), CWCD was seen as a way of evaluating whether species interactions, and more specifically interspecific competition, were responsible for determining patterns of species coexistence and distribution. This is because CWCD was expected when competition was strong and the community was at equilibrium. Alternatively, if coexisting species were more similar than expected by chance alone, then it was suggested that competition was weak, or the community was not at equilibrium.

While early studies often purported to find examples of CWCD (Hutchinson 1959; Schoener 1965; Pianka 1973), several issues arose. First, despite the examples in support of CWCD, there were still many instances where locally coexisting species were actually more similar to one another (e.g., Cody 1973). Second, better null models were needed to statistically determine whether species were more different from one another than expected by chance alone (Strong et al. 1979; Simberloff and Boecklen 1981; but see Losos et al. 1989). The issue of appropriate null models is still being debated in the literature (Brown et al. 2000; Stone et al. 2000), and we will not discuss this any further here.

There appear to be only a few noncontroversial studies that suggest that CWCD occurs (Bowers and Brown 1982; Schoener 1984; Dayan et al. 1990) whereas there are many that suggest the coexistence of similar species. Does this mean that in general, interspecific interactions do not determine species distributions? Are communities instead limited by biogeographic, historical, and/or abiotic factors? If species are more similar than expected by chance, and CWCD does not occur as a community approaches equilibrium, then it is reasonable to conclude that interspecific interactions are not important in determining these species distributions. But in fact the issue is more complicated; in general, the supposition that interspecific interactions should always lead to CWCD is incorrect (Leibold 1998).

As the basic species-sorting process illustrates (fig. 7.1), the uninvasible species pairs are actually the species whose requirements (ZNGIs) are most similar to one another, whereas the species that have more distinct ZNGIs are always invasible by a species that is more similar to one species or the other. This is the exact opposite of the prediction of CWCD! In fact, several ecological and evolutionary theoretical models have come to a similar conclusion (May and MacArthur 1972; Abrams 1975; Roughgarden 1976; Turelli 1978, 1981; Taper and Case 1985; Case and Taper 2000). That is, at equilibrium strong interspecific interactions, including competition, can actually lead to community-wide convergence.

However, this discussion relates only to one component of a species' niche: its requirements. The expectations for when CWCD should occur will depend critically on whether traits related to a species' requirements or impacts are being considered. Species that have divergent impact vectors are expected to coexist over a broader range of envi-

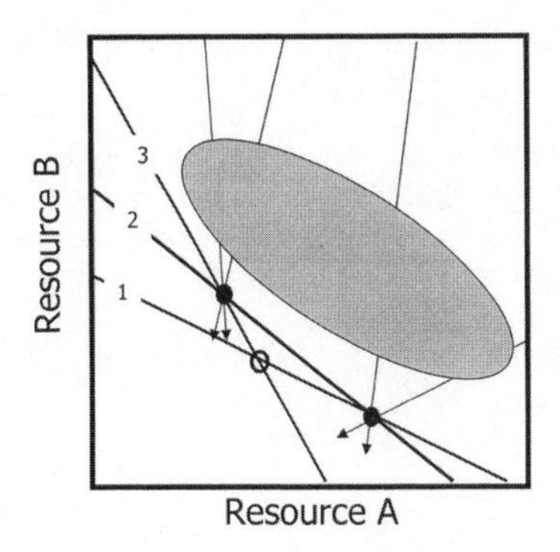

Fig. 7.5. Communitywide character displacement for requirements and impacts. Three species compete for two substitutable resources. The uninvasible equilibria (filled circles) are between species pairs that are most similar in their requirements (1–2 and 2–3), while the most dissimilar species pair (1–3) is invasible. However, the species pair that is most dissimilar in its impacts (1–2) will coexist more frequently than the species pair with the most similar impacts (1–3) over a broader range of resource supply (shaded oval). Redrawn from Leibold 1998 with permission.

ronmental conditions than species with more similar impact vectors (fig. 7.5). Thus, we would expect that species with divergent impacts would be found at many localities within a heterogeneous region, while species with convergent impacts would be found only in a narrow range of localities. Further, because species found only in such a range might be more vulnerable to local or regional extinction, we might expect to find species with more divergent impacts to coexist more frequently than expected by chance.

While we expect CWCD in the impact component of species' niches, but not in their requirements, there could also be a tension between the two effects. Many of the parameters that go into the derivation of a species' ZNGI (requirement) also determine its impact vectors (e.g., resource capture rate; see appendix to chapter 2). Thus, the traits measured by an empiricist to determine the similarity of coexisting species may jointly influence both requirements and impacts. For example, a given species' body size may influence both the level of resource on which it can survive (i.e., its requirements) as well as its impacts on that resource. Thus, a more exact evaluation of the relationship between similarity and coexistence requires measuring these two niche components separately. As far as we know, no such data are currently available.

Similarity in previous studies has almost always been evaluated on indirect or incomplete measures of niche components. There are three common methods by which similarity of coexisting species has been determined. First, similarity has been determined by the species/genus ratio (Elton 1927; Williams 1964; Ricklefs 1987), which assumed that closely related species are more similar in niche components than less closely related species. We expect that coexisting congeners, which likely have similar requirements, should be divergent in their impacts. Second, ecologists have evaluated similarity by measurements of species' morphological traits, such as body size (Hutchinson 1959; Dia-

Table 7.1. Summary of evidence for and against community-wide character displacement.

Criterion	Cases supporting CWCD	Cases not significant	Cases indicating coexistence of similar species
Species: genus ratios[1]	0	124	56
Body size ratios[2]	7	13	1
Resource utilization overlap[3]	27	13	16

[1] Summarized from Simberloff 1970.

[2] Summarized from Gotelli and Graves 1996.

[3] Summarized from Simberloff and Boeklen 1981.

mond 1975; Bowers and Brown 1982) or aspects of morphology related to resource use, such as bill size in birds or dentition in mammals (Van Valen 1965; Karr and James 1975; Dayan et al. 1990; Dayan and Simberloff 1994), or possibly even to aspects that enhance avoidance of predators, such as wing size (Schoener 1984). Species with similar morphology are expected to have similar niches; however, it is hard to disentangle the contribution of these traits to requirement and impact components. Third, similarity has been determined by measures of niche overlap, which are usually based on diet or habitat use (e.g., Pianka 1973); this could also affect both requirement and impact components.

Table 7.1 shows a compilation of results from some empirical studies examining limiting similarity and CWCD using each of these lines of evidence (taken from the summaries in Simberloff 1970; Simberloff and Boecklen 1981; Gotelli and Graves 1996). Taken together, the results seem all over the board. Sometimes CWCD is strong, at other times species are actually more similar than expected by chance. However, when broken down into the different methodologies, we can begin to make some headway.

In studies using the species-to-genus ratio, there are typically more congeneric species that coexist locally than would be expected from chance. In conventional niche theory, we might have expected the reverse, because species within the same genus should be competitively similar. Furthermore, if speciation were allopatric, we might expect that members of the same genus would occur in different localities simply based on their evolutionary history. However, this was not the case. Instead, the data might indicate that species within the same genus are relatively similar in their requirements but divergent in their impacts (recent analyses, e.g., Webb 2000, Enquist et al. 2002, have reached conclusions similar to ours using more sophisticated analyses).

The other two measures used to explore CWCD, morphological and niche overlap measures, provide mixed results. In some cases, deviations from the random were in the direction of CWCD (by at least some null model; again we are trying to think in broad strokes rather than get caught up in the technical issues of the question). In other cases, the deviations were in the opposite direction, and the traits of coexisting species were more similar to one another than expected by chance. Recall that for these two measures, both requirement and impact components are likely to be involved. Thus, we cannot discern which traits are more closely aligned with requirements and which are more closely aligned with impacts.

7.4. Concluding Remarks

An intriguing conclusion that follows from our discussion about CWCD is that similar species might often coexist in local communities. Taking this idea further might predict that species with identical niches should often coexist as hypothesized by Hubbell (2001) due to other mechanisms (see also Wang et al. 2002). If species sorting leads to local coexistence of fairly similar species, we might imagine that selection on these species might favor further convergence in species characters. Such convergence will be a delicate process because it must happen without having either species benefit too much relative to the other (otherwise exclusion might ensue). Further, convergence of species may be affected by different overall constraints (e.g., design constraints). This would indicate that convergence might be more likely among closely related species that would be less subject to those sorts of constraints.

The process of species sorting also has important implication for thinking about other community-level questions such as succession, community assembly, and patterns of biodiversity. We examine these in the following chapter. In addition, the species sorting process has important implications for thinking about the roles of species in ecosystems, which we discuss in chapter 9, and the evolution of species traits, which we explore in chapter 10.

7.5. Summary

1) Often, complex communities can be well understood when they are broken down into functional groups of species and a few environmental factors.
2) Species sorting describes how several species that show appropriate trade-offs form distinct local communities through a process of in-

vasion and interaction. Species sorting reduces the number of possible pairs of coexisting species greatly from the total number of pairwise combinations.

3) Species sorting directly determines how species are distributed along environmental gradients.
4) The distribution of species along a gradient will be in either nested subsets or gradient replacements, depending on the community module considered, the resource ratios, and the heterogeneity of resource supply.
5) Coexisting species should be more similar in their requirement components and more divergent in their impact components than expected by chance. Neither morphological traits nor niche overlap measured by empiricists can fully evaluate these predictions, because such measurements are part of both the requirement and impact components.

CHAPTER EIGHT

COMMUNITY SUCCESSION, ASSEMBLY, AND BIODIVERSITY

In this chapter, we use the species sorting process discussed in chapter 7 to illuminate several processes of interest to community ecologists. Of the many topics we could choose from, we specifically discuss (1) how species composition varies among sites through succession and assembly and (2) how patterns of diversity respond to important environmental gradients such as disturbance and productivity.

8.1. Community Succession and Community Assembly

The concepts of community succession and community assembly are closely related (Young et al. 2001); some consider them identical (e.g., McIntosh 1985; Pianka 1999). Both succession and assembly ask what happens when you start a community from scratch; however, their foci are different. *Community succession* focuses on the trajectory of species replacements following the large-scale disturbance (or creation) of a habitat. *Community assembly* focuses more on the end state of species diversity and composition in a community following the creation of open space. Ideas of community succession often assume that there is a single climax (or equilibrium) community toward which the successional processes will move if given enough time. Ideas of community assembly allow for the possibility that a single habitat can attain communities of differing composition or diversity (i.e., alternative stable equilibria) as a result of differential history. However, there is considerable convergence between these approaches. In this section, we first focus on the mechanisms of the successional trajectory (species compositional change), assuming for simplicity that there is a single equilibrium. Then, we focus explicitly on if and when alternative stable equilibria are expected to exist due to differential historical processes.

Community Succession

Since the seminal works of Cowles (1899), Clements (e.g., 1919, 1936), and Gleason (e.g., 1917, 1926, 1927), succession has held a prominent place in the study of community ecology. Generally, the concept of succession is applied to plant species, but similar ideas can be used for animal species. From the time a given community is established, its species composition goes through a series of stages, the last of which is generally thought to be the climax community. Graphical models of the niche framework can lend considerable insight into the process and patterns of succession. Here we discuss a subset of potential mechanisms of succession with strong empirical support: competition-colonization trade-offs, specialization on open spaces, and facilitation. In this section, we draw heavily on the previous work by Tilman (1982, 1985, 1988, 1990, 1994), as well as Pacala and Rees (1998).

Competition-colonization trade-offs. This mechanism assumes that there is a trade-off among species in their ability to colonize new habitats and in their ability to consume resources and compete once in a habitat (Tilman 1990, 1994; Pacala and Rees 1998; Yu and Wilson 2001). In this scenario, when a new habitat is formed or an old habitat is cleared, the first species are able to disperse long distances and find those habitats. These species, the "pioneers," grow and reproduce quite happily in the new habitat until species that are better competitors but poorer dispersers finally arrive. The climax community thus consists only of the best competitors.

One way to represent this concept of succession in terms of the niche framework is to consider two resource axes (e.g., soil nutrients and light availability for many plant communities) and classify species into two groups (fig. 8.1a): one that consists of poor resource exploiters with higher resource requirements that still trade off in their relative resource requirements for the two resources (ZNGIs with dashed lines), and one that consists of good resource exploiters with relatively low resource requirements that also trade off for the two resources (ZNGIs with solid lines). However, due to competition-colonization trade-offs, we hypothesize that this second group consists of poorer colonizers who arrive at new sites after the poor resource exploiters. We thus imagine that succession in this system would involve initial colonization by members of the first group and that as members of the second group arrive, resources will be reduced progressively, and the later species will competitively displace the earlier species (fig. 8.1a). An important feature of this dynamic is that the availability of all of the resources declines with succession. Furthermore, during this process,

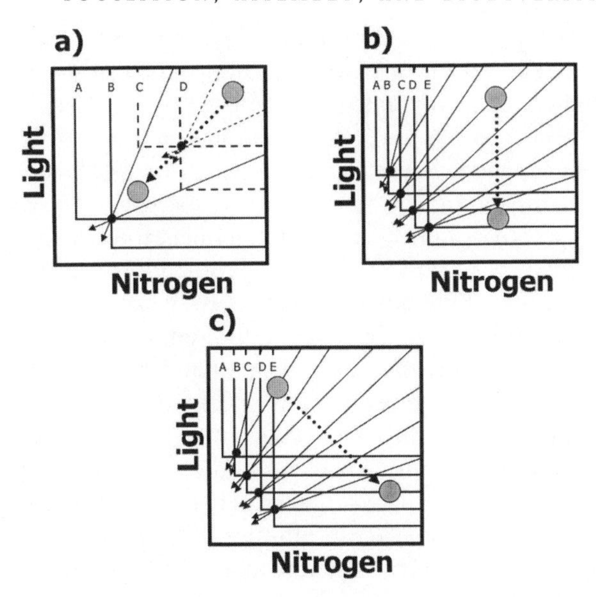

Fig. 8.1. Three mechanisms of plant community succession, each depicted by species competing along axes of nitrogen and light availability. In each case, resource supply clouds connected by arrows with dashed lines represent the temporal trajectory of succession. (a) Competition-colonization trade-offs. The first group of species (C and D) to enter a community are good colonizers (ZNGIs with dashed lines) but poor competitors, while the competitive dominants (A and B) enter the community later (ZNGIs with solid lines). (b) Specialization for open spaces. As species enter the community, the availability of superabundant light decreases. (c) Facilitation. As species enter a community, the availability of nitrogen is increased, while that of light is decreased.

all species interact negatively with each other, although the effect is asymmetric with later arrivals having disproportionately large effects on early colonists.

Specialization for open spaces. In some cases, there are species that specialize on the superabundant resources that are available when a new habitat is formed or an old one cleared (e.g., Pacala and Rees 1998). In this case, the pioneer species doesn't necessarily get to the habitat first, but instead can outcompete other species and dominate when resources are locally abundant; this case is thus different from the colonization-competition scenario. However, with time, resources associated with open spaces (e.g., light in the case of terrestrial plants) are depleted, and eventually the species that is a superior competitor when these resources are rare (i.e., the one with the lowest R^*) displaces all others and dominates. Examples of this mechanism might include a resource, such as light or some limiting nutrient that is abundant immediately following the opening or establishment of a new habitat, when succession begins, but drawn down more rapidly than other resources as other members of the community establish.

Fig. 8.1b depicts this type of successional mechanism. We assume that this mechanism involves plants competing for two resources: light, which is initially abundant but declines as succession progresses, and a soil resource (e.g., nitrogen). Think about a forest in which some major disturbance, such as a hurricane or tornado, creates a large opening awash with sunlight and assume that all of the species live somewhere nearby and are not colonization limited. If the species trade off their ability to compete for soil resources and light and the level of light is initially quite high, we would expect the species that is the least efficient light competitor but most efficient soil resource competitor to dominate. However, as individuals fill in the available space, light can be reduced disproportionately faster than nitrogen, altering the resource ratios, and become more limiting. Here then, as the resource ratios change, so does the predicted winner of interspecific competitive interactions. In contrast with the scenario involving colonization-exploitation trade-offs, only the resources associated with open space decline in availability while other resources would be predicted to increase.

Facilitation. Species that can subsist after a large disturbance do so under relatively depauperate conditions, but as they function, they influence the environmental supply rates, potentially facilitating the establishment of later organisms (Connell and Slatyer 1977). These are bioengineers (see chapter 2). For example, on a newly formed habitat such as a volcanic island, through their actions as well as through their decaying organic matter, initial colonists help to make the volcanic layer into a more habitable substrate for root growth. Similarly, nitrogen-fixing legumes can persist on low-nitrogen soils, and their presence increases soil nitrogen.

Fig. 8.1c depicts facilitation, using the case where plant species are competing for nitrogen and light. If nitrogen is exceptionally rare during the beginning of a succession, nitrogen fixers such as legumes will be strongly favored, because their R^* for nitrogen is very low. However, as these nitrogen fixers add nitrogen to the soil, they create conditions that better favor other species with higher nitrogen requirements. Further, as the level of nitrogen increases in the soil and plant abundance increases accordingly, the supply of light experienced by the plants will decrease. Hence, the supply of resources slowly changes from a high light:nitrogen ratio to a low one, and the species expected to win in competition and dominate will change through time.

Synthesis. In reality, most successional trajectories probably result from interactions among more than one of these mechanisms, all of which occur under different circumstances. For example, in newly formed habitats, "primary succession" occurs, where both colonization limitation and facilitation probably play a strong role in determining patterns of community succession. Here, the community will progress from initial dominance by colonization specialists and/or facilitating species to ultimate dominance by superior competitors. Alternatively, in habitats that are cleared of species, "secondary succession" may occur—colonization limitation may play less of a role, and succession will instead progress in a manner similar to the specialization for open patches model of succession. Nevertheless, we have ignored other possible mechanisms, such as herbivory. Including this and other factors would provide a rich array of hypotheses of successional mechanisms that can be examined and synthesized in nature.

Community Assembly

Diamond (1975) catalyzed this area of research on community assembly by discussing how species interactions (notably competition) could influence the composition of species occurring together (or not) locally (Weiher and Keddy 1999). However, Diamond's work was heavily criticized because it lacked statistical rigor (see chapter 1). Other concepts of community assembly that were based on competitive interactions and the nonrandom coexistence of different species (e.g., Brown et al. 2000) have been similarly criticized (e.g., Stone et al. 2000). In our discussion of species sorting (chapter 7), we predict a nonrandom pattern of coexistence; species are more likely to coexist with species that have adjacent ZNGIs than they are with species drawn randomly from the species pool. In a recent meta-analysis, Gotelli and McCabe (2002) found that groups of species coexisted less than expected by random chance; instead species interactions were likely to have played a strong role in driving the local coexistence of species from a diverse array of habitats and taxonomic groupings. Thus, the conjecture that species interactions and species sorting combine to determine patterns of coexistence seems to be commonly upheld.

Diamond (1975) also suggested that in some cases, even when species coexistence patterns are nonrandom, differences in the history of species colonization could lead to more than one configuration of species coexistence. In much of what we have discussed, we have predicted that communities will form quite deterministically. That is, under a given level of environmentally determined supply, we predict that a specific species or set of species will dominate, regardless of the timing in which they enter a community. The exceptions in two-species systems are the cases when a species' impacts are greatest on a factor that it finds least limiting. However, the idea that the timing of arrival of species in a community might influence the final state of that community is not novel. For example, in response to Clements's view that a single climax community would form in a given site, Gleason (1926) focused on the potential for alternative stable equilibria in the establishment of some plant communities. Similar concepts regarding the importance of species colonization order on final community outcomes are presented in Tansley's (1935) discussion of the "polyclimax" and Egler's (1954) "initial floristics model."

In the case where community assembly is strongly influenced by history, if early arriving species are able to build up high enough numbers then they will prevent establishment by later arrivals, even if those late arrivals could have survived if they had gotten there first. True alternative stable equilibria exist only when (1) each species has several chances to access the local community (e.g., there is no dispersal barrier, predators arrive after at least one of their prey species has established) and (2) any of the species could have established if it had arrived first but (3) late arriving species simply cannot establish due to the incumbent presence of prior colonizing species (Young et al. 2001). Several models have found a large number of alternative stable equilibria that result from differential community assembly histories (Gilpin and Case 1976; Drake 1990; Luh and Pimm 1993; Law and Morton 1993; Drake et al. 1999). However, these results are not necessarily robust, and other simulations have shown that only one or a few alternative equilibria exist for many types of communities (Morton et al. 1996; Law and Morton 1996; Law 1999). In all cases, however, these simulations were based on phenomenological models (e.g., Lotka-Volterra), and thus the mechanism for why alterative equilibria are or are not achieved is elusive.

Do natural communities assemble deterministically (under what are sometimes referred to as "assembly rules" sensu Diamond 1975; see also Weiher and Keddy 1999) or does history play a strong role? That is, do similar communities develop on similar sites? (McCune and Allen 1985; Jenkins and Buikema 1998). The answer from both theoretical and empirical work seems to be that it depends. Several empirical studies, many in controlled microcosms, but some in nature, suggest the presence of more than one (and sometimes many) alternative stable equilibria in environments with similar local conditions and regional species pools, differing only in the timing of the arrival of different species into the community (McCune and Allen 1985; Paine et al.

1985; Bazely and Jeffries 1986; Robinson and Dickerson 1987; Robinson and Edgemon 1988; Drake 1991; Drake et al. 1993; Lawler 1993; Samuels and Drake 1997; Petraitis and Latham 1999; Petraitis and Dudgeon 1999). On the other hand, other studies have shown that species composition had much more to do with local environmental conditions and species interactions than the order of species arrival (e.g., Neill 1975; Mitchley and Grubb 1986; Tilman et al. 1986; Sommer 1991; Grover and Lawton 1994; Weiher and Keddy 1995; Buckland and Grime 2000; Wilson et al. 2000). Furthermore, some studies have shown both types of patterns in the same study system. For example, in a study of herbaceous plant communities, Inouye and Tilman (1995) showed that replicate plots within a newly established (i.e., recently disturbed) old field, converged in community composition despite having a different initial composition. Alternatively, replicate plots within an undisturbed native savanna diverged in community composition through time as a result of different starting conditions.

What can a niche-based framework tell us about these variable patterns? First, recall that alternative stable equilibria can be predicted in niche-based models when species have greater impacts on the factor by which they are least limited or when ZNGIs intersect more than once (see chapters 2 and 5). As we discussed earlier, because plants compete for belowground resources in the soil (nutrients or water) and aboveground resources (light), species with lower requirements (R^*) for belowground resources will often have greater impacts on those resources, and species with lower requirements for light will often have greater impacts on light. This mechanism can then lead to alternative states, rather than a single climax community (Tilman 1988; Reynolds and Pacala 1993; Rees and Bergelson 1997). However, except for the specific complexities that create the opportunity for species to have the greatest impacts on resources by which they are least limited (see chapter 5), we often expect the equilibrium between species pairs to be stable. For example, in the resource competition module, if a species has low requirements for one resource type, it is likely to expend energy (either behaviorally or allocationally) acquiring the resource that it finds more limiting and thus have a greater impact on that resource (Tilman 1982; Gleeson and Tilman 1992; Vincent et al. 1996). Likewise, if a species is the best resource competitor but highly susceptible to predators, it will often have a greater impact on the predators (by providing food) than a species that is more resistant to predators (Holt et al. 1994; Leibold 1996; but see Chase, Leibold, and Simms 2000 for exceptions). Can alternative equilibria be predicted when species have their greatest impacts on the resources that most limit them?

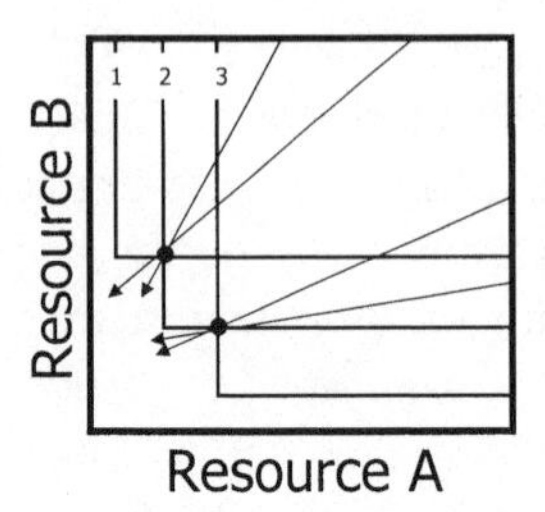

Fig. 8.2. Effects of environment and assembly history when species compete for essential resources. Here, we present three species that consume two resources. Notice that when each species has a relatively greater impact on the resource that most limits them, the final composition of the community will depend only on the local conditions of resource supply, and not the order in which species arrive into a community.

Because this will depend on both the order of arrival of the species and the mechanisms of species interactions, we discuss two community modules.

Consider the two–essential resource competition module developed by Tilman (Tilman 1982, 1988; Tilman and Pacala 1993). Fig. 8.2 shows three consumer species competing for two resources, and each consumer species can exist in the regional species pool (i.e., they have a perfect trade-off in usage of the two resources). There are two uninvasible equilibria, 1-2 and 2-3. We assume that for both equilibria, each species has greater relative impacts on the resource it finds most limiting, so that they can potentially coexist. Note that both uninvasible equilibria share species 2. Here, we need to think about the slope of the impact vector in a little more detail than we have done previously. Thus far, we have discussed the slope of the impact vector for a species only at a single point (where the ZNGIs intersect), but in fact the slope of a species' impact vector changes throughout the state space. This is because the slope of the impact results from the ratio of its impacts on each resource separately, and the ratio of the resources changes between the two potential equilibrium points (see appendix to chapter 2 for the mathematical details). The result is that the impact of species 2 will differ at the two uninvasible equilibria. In fig. 8.2, the impact of species 2 will reflect a relatively high ratio of resource B to resource A at the 1-2 equilibrium, and a high ratio of A to B at the 2-3 equilibrium. The overall slope of the impact will be shallower at the 2-3 equilibrium. The consequence of this is that the range of resource supply where we would expect coexistence for species pair 1-2 does not overlap with the range of resource supply where we would expect coexistence for species pair 2-3.

Now, imagine doing an experiment on community assembly where we give one of the species a head start and allow it to achieve its equilibrium (i.e., reduce resources to its ZNGI) prior to adding other species. In this case, if a species enters the environment first and is not the best possible competitor for that environment, future invaders will be able to displace it. The final community composition will depend on the level of resource supply, and thus the final community composition is deterministic and alternative stable equilibria will not occur.

Contrast the results discussed for resource competition with the situation in a keystone predator community module, where three species compete for a common limiting resource (R) but are also consumed by a common predator (P) (fig. 8.3a). Again, there are two potentially uninvasible equilibria where pairs of species coexist, 1-2 and 2-3. And again, species 2 is present in both equilibria. Here, we view the impact of species 2 at the different points in the state space. At the 1-2 equilibrium, species 2 has a lower P:R ratio than in the 2-3 equilibrium. This means that the slope of species 2's impact will increase as it goes from the lower to the upper equilibrium point. Because of this simple change in the slope of the impact, we can see in fig. 8.3 that there can be a range of supply points where both pairs of species at equilibria 1-2 and 2-3 can potentially coexist. But because there are only two limiting factors, only two of the three species will be able to exist at equilibrium (in the absence of any variance in resource supply). So in the range of resource supply where the coexistence realms overlap, it is uncertain which species will ultimately coexist with species 2. If species 1 enters the community first, it can drive the relative densities of resources and predators toward the range where it will coexist with species 2 and away from the range where species 3 can invade. If species 3 enters first, the roles of species 1 and 3 will be reversed. Thus, unlike the resource competition module discussed above, there are certain conditions of resource supply in a keystone predator module where we might expect alternative stable equilibria to occur, even when species impacts are greater on the resource they find most limiting, since there are more than two species present.

This case gets really interesting as more species are added. Fig. 8.3b includes five species and it's clear that species pairs have more ranges of overlap with other species pairs at the upper end of the resource gradient. That is, more and more potential alternative stable equilibria are expected. At low resource supply, only the best resource competitor (species 1) will exist; with a slight increase in resource supply, we predict a realm where species 1 and 2 will coexist. But as we continue to increase resource levels, we get more and more space where more than one pair of species is able to coexist, and which pair of species actually exists in any given environment will depend on the order in which they arrive in a community. What this means is that the number of alternative stable equilibria is predicted to increase with productivity. In addition, this prediction is consistent with the results from Chase

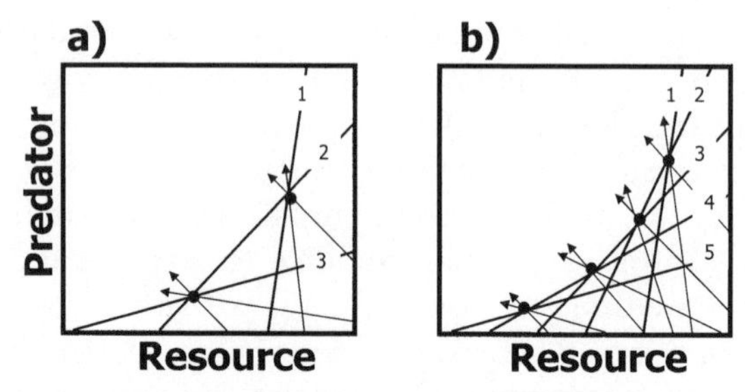

Fig. 8.3. Effects of environment and assembly history when species compete for a resource and a predator. (a) Three species that trade off and have impacts positioned such that the best resource competitor has the greatest impact on predators, and so on. Notice that there is a range of resource supplies where there is overlap between the vectors of two uninvasible species pairs. In this range, alternative stable equilibria of species pairs exist, and the final identity of species will depend on which species enters the community first. (b) Five species that show the same trade-off in ZNGIs and relative position of their impact vectors. Here, notice that with increasing levels of resources (along the *x*-axis), the ranges of overlap among coexisting pairs of species increase such that the number of potential alternative stable equilibria increases along the *x*-axis (increasing levels of resource supply).

and Leibold (2002) in surveys of pond communities; ponds with low productivity were quite similar to one another, whereas ponds with high productivity were quite divergent, suggesting a possible role for alternative stable equilibria. Note, however, that this scenario is expected only when species share both resources and predators, because when species share only resources, no such alternative equilibria are expected.

We have used the niche-based approach to address the question of whether similar communities develop on similar sites. Instead of answering "it depends," our simple models allow us to determine the conditions when we would expect community assembly to act in a very deterministic way and when invasion sequence might play a stronger role.

8.3. How Do Species Interactions Regulate Patterns in the Distribution of Biodiversity?

How many species are expected to coexist in a given locality or region? How is biodiversity distributed along environmental gradients such as productivity or disturbance? These are perhaps the most fundamental questions of community ecology, but they also have some of the most elusive answers. Indeed, recent syntheses of patterns of diversity ask many more questions than they answer (e.g., Rosenzweig 1995; Waide et al. 1999; Gaston 2000). While we by no means pretend to answer the biodiversity riddle better than our predecessors, we do suggest that a niche-based framework might shed considerable insight into these questions.

Most fundamentally, local species diversity results directly from the rules of coexistence that we discuss throughout this book. Diversity will be highest when species are predicted to coexist and lowest when one species is predicted to dominate. Furthermore, heterogeneity in the supply or relative position of species requirements and impacts will tend to increase the probabilities of species coexistence. Finally, the total number of species that can exist in a given locality will depend both on this heterogeneity and on the number of limiting factors on which they can trade off. However, obviously many more species coexist in nature than these simple models could ever predict. Thus, the greatest utility of this framework is in making qualitative predictions, particularly along environmental gradients, rather than making specific predictions about the number of species expected in a single locality.

Two environmental gradients have held very prominent roles in discussions of species diversity: (1) diversity along a disturbance gradient and (2) diversity along a productivity gradient.

Disturbance and Diversity

Disturbance is usually defined as a process that kills or seriously stresses a large proportion of the individuals of at least one species in a community (see section 8.1 above) and has been long known to have important effects on diversity. These effects have been reviewed in detail elsewhere (e.g., Sousa 1984; Pickett and White 1985; Petraitis et al. 1989; Wootton 1998; Mackey and Currie 2001), and so we only summarize the evidence and concepts here, with particular reference to the utility of a niche-based approach.

Perhaps the most famous concept of the influence of disturbance on diversity is the "intermediate disturbance hypothesis" as formalized and popularized by Connell (1978) (see also Dayton 1971; Lubchenco 1978; Sousa 1979). This hypothesis predicts that diversity will be lowest when disturbance intensity is very low or very high, and highest when disturbance intensity is intermediate. At very low levels of disturbance those species that are competitively superior drive inferior competitors extinct, while at very high levels of disturbance only those species that are superior at dealing with disturbance (e.g., pioneer or weedy species) can persist. At intermediate levels of disturbance, both groups of species are expected to persist at least in some places and at some times because of the continued flux of disturbance in the community. Here, the community is not so disturbed that only pioneer species can live, but time between disturbances is frequent enough to prevent exclusion by the competitive dominants.

There are two ways in which patterns of the intermediate disturbance hypothesis can be manifest; both depend on the spatial scale at which disturbance acts and on which species diversity is counted. First, if disturbance acts on the entire community, then habitats with intermediate disturbance rates are expected to be in the transition between dominance by one species or the other, and the species will coexist locally for a brief period of time. This is likely the sort of process observed by Sousa (1979), who found that large boulders were rarely disturbed by winter storms, small boulders were frequently disturbed, and medium-sized boulders were intermediate in their disturbance frequency; algal diversity was highest on the medium-sized boulders. Thus, disturbance maintained diversity within a single locality (in this case a boulder) by resetting the successional trajectory frequently enough to prevent exclusion by the competitive dominants, but not so frequently that only pioneer species could exist there.

Second, disturbance can maintain diversity in a region of local communities if localities are disturbed at different time intervals (e.g., Levin and Paine 1974; Paine and Levin 1981). If disturbances were frequent, then we would expect that most localities would have been disturbed relatively recently, and thus each would contain a low diversity of pioneer species. Alternatively, if disturbances were much less frequent, we might expect that most localities would have gone through the successional trajectory and contain a low diversity of only competitively dominant (climax) species. However, at intermediate levels of disturbance, the time since each locality within a region of communities was disturbed would be highly divergent. Thus, each locality might be in a different stage of succession as it transitions from pioneer to climax species. Some communities would have been disturbed relatively recently and contain pioneer species, whereas others would have been disturbed a long time ago and contain mostly climax species. Paine and Levin (1981) provide good evidence for this mechanism for the maintenance of diversity through disturbance on the rocky shores off the coast of Washington State, where the middle-intertidal zone is dominated by mussel beds, which seem to be the competitively dominant climax species. However, in exposed sites wave action can rip large areas of these mussel beds from the rocks, creating large patches of habitat available for colonization by pioneer species. At any given locality, diversity may not differ that much (although composition will

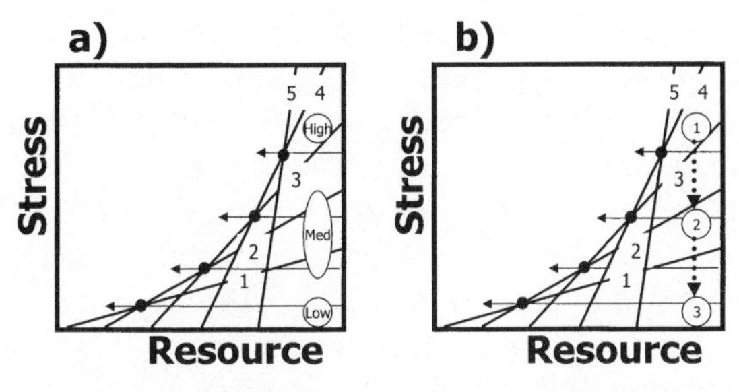

Fig. 8.4. The effects of disturbance on species diversity when species trade off in their relative ability to compete and tolerate stress (disturbance). (a) The shaded clouds represent the variance in disturbance supply in different communities with different disturbance rates. We assume here that the variance in disturbance rate is highest at intermediate levels of disturbance and lowest at low and high levels of disturbance. (b) Here, the shaded clouds represent a community at a different time following the last disturbance, ranging from the most recently disturbed (labeled 1) to least recently disturbed (labeled 3).

differ greatly), but in the entire mosaic of habitats at different successional stages diversity is high at the regional level.

Although the two distinctions of how disturbance can maintain diversity at local and regional scales may be subtle, we use the graphical framework to illustrate the importance of considering these different mechanisms. We define disturbance in the graphical framework in a way quite similar to the axis of stress that we have used in other sections in this book. Species better at dealing with disturbance are expected to be poorer resource competitors (higher R*) than are species that are more vulnerable to disturbance, such that their ZNGIs intersect. This differs significantly from Huston's (1979, 1994) model of how disturbance maintains diversity, which unrealistically assumes that all species are equally vulnerable to disturbance, but is in accord with more mathematically rigorous models of how species coexistence is altered by disturbance and stress (e.g., Chesson and Huntly 1997; Chesson 2000b).

We discuss the difference between intermediate disturbance at local and regional scales, and its consequence for diversity. We assume that five species are sorted along resource and stress axes (fig. 8.4). In the case where disturbance maintains diversity in local communities, species diversity can be highest at intermediate disturbance only if the variance in the disturbance itself is highest at intermediate disturbance (fig. 8.4a). This was likely the case in Sousa's (1979) study, since the variance in large and small boulders was very low, but was high on medium boulders, which are disturbed only by intense wave action.

In the case where disturbance maintains diversity at the regional scale, for simplicity's sake we ignore the case of very high disturbance and very low disturbance, where all localities within a region are expected to be at similar successional stages (and thus similar levels of diversity). However, at intermediate disturbance, we assume that communities can be in one of three states that relate to how long it has been since they were disturbed. We further assume that a given disturbance event is strong and favors the species that is the best pioneer. Thus, at any given point of time, each community will be in a different stage of succession and will be characterized by a different suite of species. If each local community is at a different point along the successional trajectory (labeled 1–3 in fig. 8.4b) and we were to measure the species diversity at any given locality along a gradient of disturbance, we would find a similar diversity of species at each site. But if we were to view the same group of species from a larger regional perspective, including all three communities at different points along their successional trajectory, we would see the peak in diversity at intermediate levels of disturbance because of the addition of different species that dominate each locality.

Although a variety of studies have supported the patterns predicted by the intermediate disturbance hypothesis at local and/or regional spatial scales, several studies have failed to find such evidence (see, e.g., Wootton 1998; Mackey and Currie 2001). The difference in the scales at which the response of diversity to disturbance is expected in these two mechanisms might explain some of this empirical discrepancy; this is particularly true because empiricists often measure species diversity at different spatial scales without making scale explicit. However, using general niche-based models, Wootton (1998) suggested two other general reasons why species diversity might not peak at intermediate levels of disturbance: (1) if species do not respond differentially to disturbance, then disturbance will not increase the probability of species coexistence (see also Chesson and Huntly 1997; Chesson 2000b), and (2) if species coexist by means other than the action of disturbance, then disturbance will either have no influence on species coexistence or will decrease the probability of species coexistence and uniformly reduce diversity.

Productivity and Diversity

The effect of primary productivity on patterns of diversity has enjoyed considerable interest, dogma, and debate in the literature. A variety of pattern analyses and experimental results have suggested that the relationship between productivity and diversity is often hump shaped. The hump-shaped pattern describes a situation where species

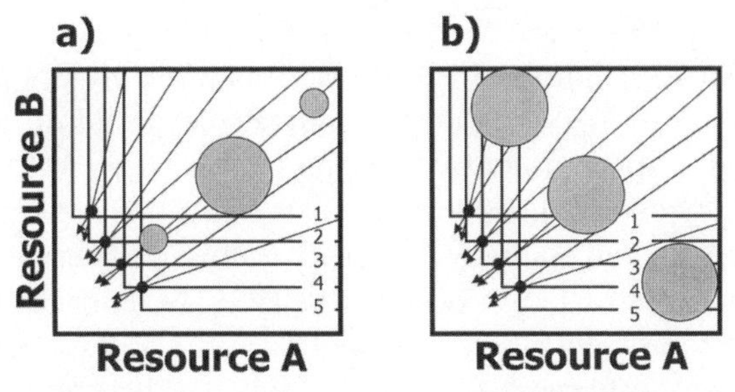

Fig. 8.5. Analyses of the relationship between primary productivity (nutrient supply) and species diversity. In each case, we assume there are five species competing for two essential resources and show how certain assumptions can lead to the often observed hump-shaped relationship between productivity and diversity. (a) Here, we assume that as the abundance of resources increases from the lower left to the upper right, the degree of heterogeneity is small at low productivity, high at intermediate productivity, and low again at high productivity. (b) Here, we assume that as the level of one limiting resource increases, the level of the other limiting resource decreases, as might be envisioned when plants compete for soil nutrients and light (redrawn from Tilman and Pacala 1993 with permission).

diversity is lowest at low productivity, peaks at intermediate levels of productivity, and declines again at high productivity.

Several conceptual models have been developed to explain this pattern (reviewed in Waide et al. 1999). We present an abbreviated version of some of these models within a niche framework, using the resource competition (Tilman 1982; Tilman and Pacala 1993) and keystone predator (Leibold 1996) modules. In the resource competition module we consider two mechanisms proposed to explain the hump-shaped patterns, the resource heterogeneity model and the resource ratio model (fig. 8.5). In each case, we consider the interactions and coexistence among five species of hypothetical plants that show perfect trade-offs in both their requirements (ZNGIs) and impacts. The resource heterogeneity model (fig. 8.5a) assumes that the heterogeneity of those resources, denoted by the diameter of the supply circle, is highest at intermediate levels of the two limiting resources (keeping ratios constant) (Tilman 1982). The resource ratio model (fig. 8.5b) assumes that species are competing for a limiting nutrient and light. If nutrients limit the biomass of plants, then increasing levels of nutrients will subsequently decrease levels of light. Thus, with increasing productivity, the resource ratios will change from high levels of light and low levels of nutrients to high levels of nutrients and low levels of light (fig. 8.5b) (Tilman and Pacala 1993).

Tilman and Pacala (1993) presented a number of cases where patterns of plant species diversity supported the hump-shaped productivity-diversity pattern, and Rosenzweig (Rosenzweig 1995; Rosenzweig and Abramsky 1993) included both studies of plants and animals to conclude that the hump-shaped pattern was probably ubiquitous (see also Huston 1994, 1999). On the contrary, while hump-shaped patterns are found in a variety of ecosystem types (Leibold 1999; Waide et al. 1999; Gross et al. 2000; Dodson et al. 2000), they are by no means universal, and a variety of other patterns are seen in both plants and animals (Abrams 1995; Waide et al. 1999; Mittelbach et al. 2001).

Abrams (1988, 1995) has critiqued Tilman's models primarily because different patterns are predicted when other assumptions are made. Of course, determining the appropriate assumptions is ultimately an empirical question. Nevertheless, with some simple manipulations of the graphical models discussed above, you can easily see how different assumptions might lead to different productivity-diversity relationships within the resource competition module. We start with the assumption that resource supply will always be above the level where at least one species can exist—if this were not the case, species diversity would always initially increase from zero to one or two, and then either continue to increase, stay the same, or decrease. Further, different assumptions of the relationship between resource supply and productivity can lead to several different predictions of productivity-diversity relationships, including positive, negative, hump shaped, bimodal, and no relationships (see also Abrams 1995).

While the resource competition module has been the most popular model for exploring productivity-diversity relationships, it is not the only one, nor is it necessarily the most realistic. Rosenzweig (1995) and Leibold (1999) have reviewed a number of other hypotheses that have been used to explain the hump-shaped productivity-diversity pattern. In particular, the keystone predator module (Leibold 1996), though it has received less attention than the resource competition module, might be better able to predict the hump-shaped productivity-diversity pattern in some types of empirical systems (Leibold 1999).

In the keystone predator module (fig. 8.6a), as resource supply increases along the *x*-axis, so does productivity. In its simplest form, with two species that compete for a common predator and a common resource, we predict that at the lowest levels of productivity, only the superior resource competitor exists alone, the species coexist at intermediate levels of productivity, and the better-defended species exists alone at the highest levels of productivity. This is a hump-shaped productivity-diversity pattern. Of course, in this case, species diversity only went from 1 to 2 to 1 again. However, Leibold (1996) showed how the incorporation of several species, with a moderate amount of environmental heterogeneity in resource supply, could

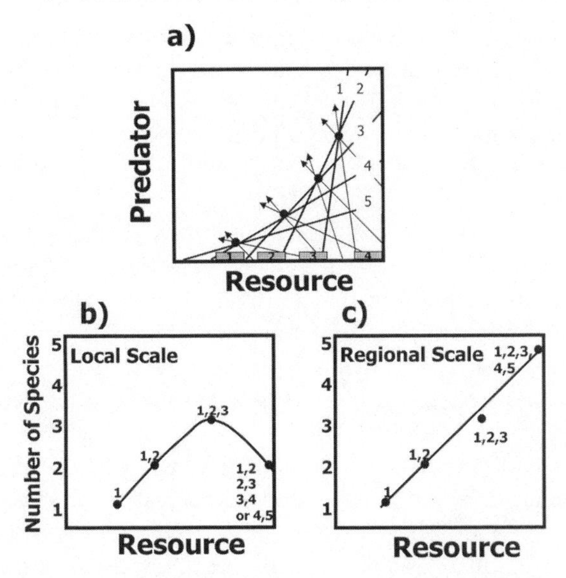

Fig. 8.6. Using the niche framework to predict a scale-dependent diversity relationship. Here, we assume that species trade off along axes of predators and resources. (a) ZNGIs, impacts, and ranges of resource supply for patterns of coexistence for five species. Boxes with numbers refer to four qualitatively different levels of resource availability (i.e., primary productivity). (b) Species diversity along the productivity gradient predicted at the scale of an individual locality. (c) Species diversity along the productivity gradient predicted at the scale of the region. This prediction is due to the predicted increase in the number of alternative stable equilibria with increasing productivity (see also fig. 8.3), such that most of the species are expected to exist in at least one locality within the region.

lead to a realistically diverse group of species that showed a hump-shaped productivity-diversity relationship. Furthermore, Leibold (1999) showed that the hump-shaped relationship between productivity and plankton diversity found in a large region of ponds most likely resulted from mechanisms associated with the keystone predator module.

However, recall that in the keystone predator module, we predict that as the levels of resource supply increase, more and more alternative stable equilibria of pairs of species will occur (fig. 8.6a). The hump-shaped productivity-diversity relationship occurs in the keystone predator model only when species diversity is viewed at the local scale: that is, when productivity and diversity are measured within any one local community (fig. 8.6b). However, what would happen if we were to look at species diversity among several local communities within regions of similar productivity? For example, a low-productivity desert consists of several local communities, as does a high-productivity forest. If we went to any one site within the desert or forest, we might expect to find a low species diversity because of the hump-shaped relationship. However, the keystone predator module predicts that at high productivity, different localities within a region reside in alternative stable community states. Using Whittaker's (1972) terminology, the turnover or change in species composition from site to site within a region, due to alternative states or other types of heterogeneity, is called β-diversity.

Why should we care about β-diversity? Whittaker's full diversity equation shows that the total diversity of species in a region (γ-diversity) will be the product of species diversity at any given locality (α-diversity) and the compositional turnover of species among localities (β-diversity). That is, $\gamma = \alpha * \beta$ (qualitatively similar results are obtained if this relationship is additive [$\gamma = \alpha + \beta$] as preferred by Lande 1996, Rosenzweig 1999, and Loreau 2000a). Thus, while the keystone predator module predicts a hump-shaped productivity-diversity relationship when diversity is measured on a local scale, it also predicts an increasing relationship between productivity and β-diversity. This means that regional γ-diversity will actually increase monotonically with productivity (fig. 8.6c). This is because at the highest levels of productivity, in any given locality, species diversity is low, but the identity of those species will be different from the identity of species in any other locality because of the predicted alternative stable states, and thus γ-diversity will be highest at high productivity. Thus, the keystone predator module actually predicts a scale-dependent pattern of the productivity-diversity relationship. No such scale dependence is predicted in the resource competition module described above, because it does not predict increasing levels of β-diversity with productivity.

Do we see this sort of scale dependence in nature? Yes. It has long been noted that at relatively large spatial scales that incorporate several localities, diversity typically rises with rising productivity (e.g., Brown 1981; Currie 1991). The hump-shaped pattern is usually observed when species diversity is taken from a single locality (Tilman and Pacala 1993; Rosenzweig 1995). Gross et al. (2000) and Mittelbach et al. (2001) explicitly looked for these sorts of scale-dependent relationships and found suggestive evidence for them. Also, in our surveys of freshwater ponds in Michigan (Chase and Leibold 2002), when species diversity was viewed from within any local pond, we found a hump-shaped pattern (fig. 8.7a), whereas when the same data were viewed from a regional scale, we found a monotonically increasing pattern (fig. 8.7b). Further, we found that this pattern resulted because ponds within regions of higher productivity were more divergent from one another in species composition (i.e., they had higher β-diversity) than were ponds from lower-productivity regions.

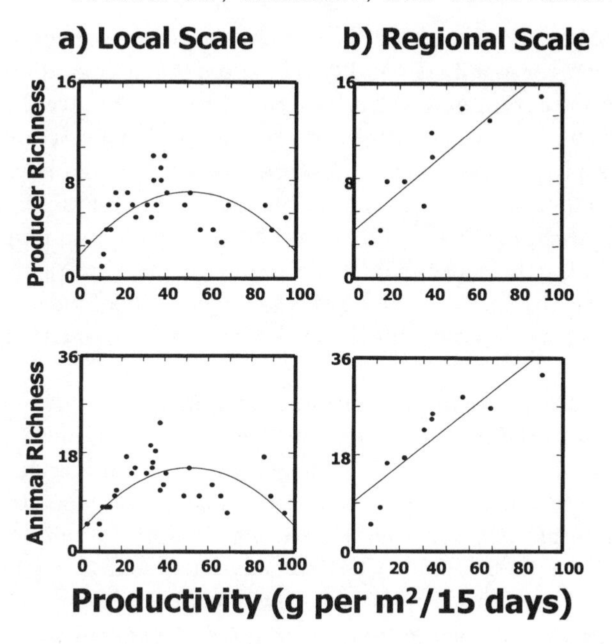

Fig. 8.7. Data from a survey of pond plants and animals showing the scale-dependent productivity-diversity relationship. (a) Diversity when viewed at the local scale (a single pond). (b) Diversity when viewed at the regional scale (the addition of multiple ponds). Redrawn from Chase and Leibold 2002 with permission.

Of course, this does not mean that the mechanisms of the keystone predator module are exactly those mechanisms that determine these scale-dependent patterns. If increased heterogeneity accompanies increased productivity, this too could cause an increase in β-diversity. Thus, local sites within higher-productivity regions could be more heterogeneous than local sites within lower-productivity regions. Alternatively, higher productivity sites could experience higher temporal turnover of species composition.

8.4. Concluding Remarks

We have left many interesting and important topics that may benefit from a niche-based perspective out of this chapter and the previous one. For example, the presence of exotic and invasive species has been implicated as a serious agent in the global loss of biodiversity. And yet the theory of exotic species is still in its infancy in many respects. What makes certain species invasive? What makes certain communities invasible? A niche-based perspective should prove quite useful in this context (see Shea and Chesson 2002). Regions with fewer species might have more divergent ZNGIs and thus be more invasible by species from more diverse communities that have traits that would give them more intermediate ZNGIs. Likewise, we predict that local communities whose supply points have been anthropogenically altered beyond their historical range will favor exotic species from a different regional pool. For example, Miller et al. (2002) found that local communities of protists inhabiting the water-filled leaves of pitcher plants were more likely to be invaded by species from outside the regional pool when resources or predators were augmented than when environmental conditions were unaltered.

Community ecology, perhaps more than other subfields of ecology, seems to be mired in contingency, the answer to most questions being "it depends." This is frustrating for everyone from undergraduates to seasoned researchers. So, community ecology lacks generality in the sense that a physicist or even a molecular biologist might define the term. Lawton (1999, 2000) has recently suggested that community ecology is so wrought with contingency that perhaps researchers should abandon it for the greener pastures of generality that can be seen at higher scales, such as macroecology (see also Maurer 1999). However, the niche-based approach provides a way to predict when and how systems should or should not differ. If we find different productivity-diversity relationships in different types of communities, we can begin to know why. Thus, we feel that the true generality of community ecology lies not with universal truths, but rather with an informed set of tools that allows us to understand, predict, and synthesize the variability that is seen among communities in both pattern and process.

8.5. Summary

1) Competition-colonization trade-offs, specialization for open spaces, and facilitation are three very different mechanisms by which succession can take place. The niche concept can be used to place common currency on each of these mechanisms and predict how succession should proceed following a disturbance.
2) Community assembly can be highly deterministic if the environmental conditions and interspecific interactions determine which species exist in a local community, or highly stochastic if historical differences in colonization order lead to alternative stable equilibria. Two modules, resource competition and keystone predation are

used to discuss when deterministic or stochastic factors are more likely to determine community composition.

3) Insights from the niche framework can be used to make predictions about patterns of species diversity along environmental gradients; we focus on disturbance and productivity.
4) The intermediate disturbance hypothesis predicts highest species diversity at intermediate levels of disturbance. The resource-stress community module predicts that generally the intermediate disturbance hypothesis is more likely to be supported when species diversity is viewed at larger spatial scales, incorporating many local communities.
5) Two niche-based modules, resource competition and keystone predation, can predict hump-shaped patterns between primary productivity and species diversity under certain assumptions.
6) The keystone predation module, but not the resource-competition module, predicts that with increasing productivity the number of alternative stable states should increase. From this, we predict that in this module, the relationship between productivity and diversity should be scale dependent: hump-shaped at local scales and increasing at regional scales. Such scale-dependent patterns are often observed in natural communities.

CHAPTER NINE
NICHE RELATIONS WITHIN ECOSYSTEMS

Ecosystems are defined as interactive systems that involve communities of species along with their abiotic environment (Tansley 1935; Odum 1951). From this definition, ecosystems should have intimate links with community ecology. Further, given the strong connection between the niche concept and patterns in community ecology, we might expect the niche to play a key role in ecosystems ecology. Early on, it did. Elton (1927) and Tansley (1935), who played important roles in the development of the niche concept (see chapter 1), were also instrumental in developing ecosystems ecology as we now know it. However, today the niche framework is seldom used to conceptualize ecosystems ecology. This is because ecosystem ecologists who focus on flows of energy and nutrients at large scales often downplay the role of species within ecosystems. Nevertheless, recently there has been a redefined focus on the interface between community and ecosystems ecology that is unearthing some interesting and important phenomena and will continue to do so (e.g., Jones and Lawton 1995; Kinzig et al. 2002).

In our separation of the niche concept into two components, requirements and impacts (see chapter 2; see also Leibold 1995), the second component describes the role of the species in the ecosystem and illustrates the potential for the niche concept to contribute to ecosystem-level questions. Recent empirical evidence has shown the potential importance of species within communities on ecosystem-level processes. For example, within communities, there are often particular species with very strong, perhaps disproportionate, effects on ecosystems-level processes (e.g., Power et al. 1996; Berlow et al. 1999). In addition, several recent studies have found a positive relationship between species diversity (or functional group diversity) and ecosystems function (reviewed in Naeem et al. 2000, Loreau 2000a, Loreau et al. 2001, and Kinzig et al. 2002; but see Huston et al. 2000 and Wardle et al. 2000), although species composition may play at least a big a role as species diversity per se in some cases (Hooper and Vitousek 1997; Downing and Leibold 2002). In addition, most of these studies have only considered competitive interactions within a single trophic level, whereas among–trophic level effects of species diversity may be considerably strong (Naeem and Li 1997; Downing and Leibold 2002; Paine 2002).

Clearly, there are important ways in which ecosystem attributes are related to community structure. For example, in chapter 8 we discussed how attributes of the ecosystem, such as its primary productivity, resource ratios, and disturbance rates, influenced community attributes such as species composition and diversity. This chapter addresses whether the reciprocal effects of the community on the ecosystem are also important. We build upon a recent upsurge of empirical, and to a lesser degree theoretical, studies exploring the role of biodiversity on the functioning of ecosystems. Although ecosystem functioning is often vaguely defined, in general it has primarily been applied to primary productivity or standing biomass. Other types of ecosystem functions include nutrient cycling and decomposition rates, but we focus less on these.

In our discussion, we follow Loreau (2000b) and discuss in turn the short-term effects of biodiversity on ecosystem function and the longer-term effects of biodiversity on ecosystem stability in the face of perturbations. Then we explore how the process of species sorting, as discussed in chapter 7, might influence the results of ecosystem perturbations in the short and long terms.

9.1. Short- and Long-term Effects of Biodiversity on Ecosystem-level Properties

When biodiversity is treated as an independent variable, its effects on ecosystem-level processes have given mixed results; sometimes the effects are quite strong, and at other times they are weak or not detectable (reviewed in Naeem et al. 2000; Wardle et al. 2000; Loreau et al. 2001; Kinzig et al. 2002). The relationship between species diversity and ecosystem properties has taken a particularly prominent role in recent community and ecosystems ecology for two reasons.

First, it is obviously important in guiding environmental policy. If biodiversity, or the components associated with biodiversity (such as species differential roles within the ecosystem), plays a significant role in the functioning of ecosystems, then national and international policies on conservation of biodiversity take on added significance (Chapin et al. 2000). This is especially true in the light of recent studies suggesting that many of the world's ecosystems provide valuable services to humans that can be quantified in economic terms (Costanza et al. 1996; Daily et al. 2000; Balvanera et al. 2001; Balmford et al. 2002).

Second, and perhaps more importantly, research on the interface between community and ecosystems ecology is revealing that our scientific understandings of the roles of species (in terms of both absolute numbers and particular species) in ecosystems is very incomplete. Questions related to these issues point at important lacunae in our understanding, knowledge, and insights about fundamental aspects of ecology.

Several recent theoretical studies have used the niche concept in understanding the role of diversity in ecosystem processes (Tilman et al. 1997a; Tilman 1999; Loreau 2000b; Petchey 2000; Bond and Chase 2002; Chesson et al. 2002; Holt and Loreau 2002; Kinzig and Pacala 2002). We use the niche-based framework to overview the essential concepts in this work and put it in a more synthetic context. For short-term effects, we ask how the biodiversity within a local site influences the functioning of that ecosystem—here, ecosystem functioning includes variables like primary productivity, efficiency of nutrient use, decomposition rates, and turnover. For long-term effects, we ask how the biodiversity within a local site influences the long-term stability of that ecosystem—here, stability includes factors such as variability (the degree of variation around a mean), resilience (the speed with which a system returns to equilibrium following a perturbation), and resistance (the degree to which a system is moved away from equilibrium by a perturbation) (Pimm 1991). In both cases, we are treating biodiversity as an independent variable. That is, we are asking how variation in biodiversity, independent of variation in any other factor, influences these variables. Obviously, however, there will be reciprocal effects; productivity affects biodiversity, and biodiversity affects productivity. These issues are harder to deal with and have just begun to be addressed.

Short-term Effects

Some of the major hypotheses about the role of species and biodiversity on ecosystem functioning can be illustrated and testable predictions can be made using a niche-based approach (see also Tilman 1999):

(1) The *niche complementarity effect* (Darwin 1859; Elton 1927; Tilman, Lehman, and Thomson 1997; Loreau 1998). If species trade off in their

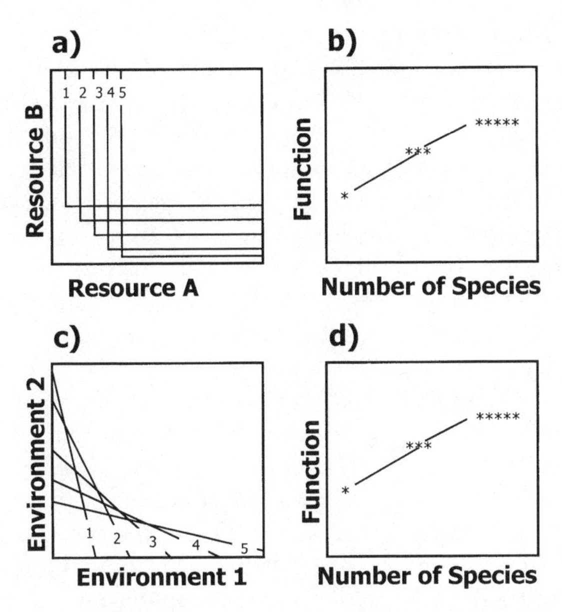

Fig. 9.1. Two niche-based hypotheses for the positive relationship between diversity and ecosystem function. Left panels depict the niche-based model, and right panels depict the hypothetical relationship between species diversity (at three arbitrary levels) and some measure of function. (a, b) Each species utilizes resource types differently. Because the species have complementary resource usage, each has an equal effect on ecosystem function. (c, d) Each species responds differently to two different environments. The species most efficient in one environment is least efficient in the other. Thus, having more species present increases the overall function when the environment fluctuates.

use of different ecosystem variables, then greater diversity will result in greater overall functioning. This effect can be illustrated in the resource competition module with two resource types, A and B. In a regional species pool that has passed through the species sorting filter, no single species will be simultaneously able to utilize both resources more effectively than any other species. Thus, along two resource axes, the sequential addition of species will cause those species that can coexist to utilize the resources more and more efficiently (fig. 9.1a). Because of this, the overall functioning of the ecosystem will increase with increasing species diversity (fig. 9.1b). For example, when more species are present, they are able to use the available resources more effectively and generate higher productivity (or biomass); this relationship has been long recognized in agriculture (e.g., Vandermeer 1992).

(2) The *insurance effect* (McNaughton 1977; Walker 1992; Lawton and Brown 1993; Naeem and Li 1997; Yachi and Loreau 1999) (sometimes also called the portfolio effect). If species vary in how effective they

are under different environmental conditions and environmental conditions vary temporally, then at high diversity the most efficient species for the current environmental condition is more likely to be present (see also Chesson 2000b; Chesson et al. 2002). We illustrate this idea in a way quite analogous to the niche complementarity effect above, but we consider resources in two different types of environments that can vary temporally as our axes. Further, we assume that species trade off such that the species that is best in one environment is the worst in the other environment (fig. 9.1c). If the environment were constant, a single species (the one with the lowest R^* in the single environment) would be able to maintain the highest level of ecosystem functioning. Because this species is expected to be worse at dealing with the alternative environment, it will perform much worse if the environmental conditions change. However, as species diversity increases, other species will be present that are more adept at dealing with the alternative environmental conditions, and they can compensate to maintain overall ecosystem functioning as the environmental conditions vary (fig. 9.1d). A similar effect may also occur when environments vary spatially rather than temporally (Bond and Chase 2002).

(3) The *sampling effect* (Huston 1997; Wardle 1999). This simple probability argument has been put forth as an alternative to the above niche-based arguments for the experimental effects of diversity on ecosystem functioning variables. If species vary in how effective they are in maintaining ecosystem functioning, then at high diversity those effective species have a greater probability of being present in the experimental treatment. Even though this argument is a statistical one, it is a real ecological process that could lead to an overall positive relationship between diversity and ecosystem functioning. This idea can be illustrated in a niche-based perspective in a very intuitive way. Consider a group of species in a regional species pool, each of which has a different ability to consume resources (R^*). Assume that several species can coexist in a given locality with multiple limiting factors and/or spatio-temporal heterogeneity. However, several of these species that can exist in a given locality will not be the best under the particular environmental conditions within that locality. That is, each species' ability to maintain function in the ecosystem can be related to its ability to reduce resource levels. If species from the pool were randomly put into or taken out of a locality, the probability that the species with the greatest ability to reduce resource levels will be present in that community will be X/N, where X is the number of species in a given locality and N is the number of species in the regional pool. As $X \rightarrow N$, the probability that the best species for a given locality will be present increases, and

thus the overall ecosystem function should increase with increasing numbers of species.

In reality, these three processes are likely to act in concert in natural ecosystems but may be expressed a different temporal (or spatial) scales. For example, in a long-term experiment manipulating plant diversity, Tilman and colleagues initially found good evidence for patterns similar to the sampling effect, whereas when the experiment was allowed to run for a longer period of time and the species drew down resources, patterns more aligned with niche complementarity became evident (Tilman et al. 2001). We could even imagine that as this experiment continues and the environment varies on a longer time scale, the insurance effect may also begin to play a role. The niche-based framework allows us to view these processes using a common currency and to make explicit predictions. In the future, we could see much progress being made in this arena by explicitly considering the role of the niche on patterns of natural species sorting and community assembly and their consequent effects on patterns of diversity, species composition, and ecosystem functioning.

Long-term Effects

A somewhat distinct matter that is nevertheless intimately related to the relationship between diversity and ecosystem functioning is the relationship between diversity and ecosystem stability. MacArthur (1955) was one of the first to formalize a rather straightforward hypothesis: more speciose ecosystems should be more stable than less speciose ecosystems. That is, an ecosystem with a high diversity is less likely to be influenced by an environmental perturbation, at least at the coarse level of ecosystem functioning, than an ecosystem with lower species diversity. This is because in more diverse communities, there is a higher likelihood that if one species were adversely affected by the perturbation, there would be others to take its place. In some ways, the fact that more diverse communities should be more stable is an inevitability of statistical averaging (Doak et al. 1998), but this process is also important ecologically (Tilman et al. 1998) and depends critically on the definition of stability that is considered.

May (1974b), using relatively simple linear Lotka-Volterra equations, made the startling and still controversial (Loreau 2000b) prediction that communities actually became less stable as more and more species were added. However, his conclusions depended on the form of models he used, as well as his particularly limited definition of stability.

For example, in May's approach, a community is said to be unstable if any one species in the community cannot persist. Further, May randomly assigned traits (interaction coefficients) to species in a community, rather than allowing them to sort according to their traits. Given a simple probability argument, if species are randomly added into a community, the addition of more species would increase the probability that any one of them would not be able to persist, and we would say that the community was unstable. In reality, however, this community is just not well sorted; a more diverse well-sorted community might be more stable than a less diverse one. Indeed, using a more complex approach, McCann et al. (1998) (see also McCann 2000) showed theoretically that when species interactions were more naturally dispersed, more diverse communities were more stable.

There are now more precise words to describe various aspects of how ecosystems respond to perturbations (see Pimm 1991; Loreau 2000b). *Resistance* refers to some measure of how hard it is to push an ecosystem from its preperturbation state. *Resilience* refers to how quickly an ecosystem returns to the preperturbation state. *Variability* (also referred to as *predictability*) refers to how the functions of an ecosystem vary through time (McGrady-Steed et al. 1997), which is conceptually the inverse of Naeem and Li's (1997) concept of reliability. May's conclusions about stability are most closely related to resilience and ignore the other two possibilities. Neubert and Caswell (1997) show that there may be very little correspondence between this type of stability and other components of variability in ecosystems.

Empirically, research on this question was reinvigorated by Tilman and Downing's (1994) observations that following a drought, experimental plots with higher diversity tended to rebound back to their predrought levels of biomass much quicker than did plots with lower diversity. However, this result was confounded because lower-diversity plots were those that had received nutrients for several years and this nutrient addition caused plot diversity to decline (Tilman 1996, 1999). More recently, when diversity has been experimentally manipulated, studies in both aquatic and terrestrial ecosystems have tended to show that more diverse ecosystems were more stable (by several different measurements) than less diverse ecosystems (e.g., McGrady-Steed et al. 1997; Naeem and Li 1997; Petchey et al. 1999; Downing and Leibold 2002). While the niche-based approach cannot delve into all of the issues of stability, and we leave the issues of population fluctuations and variability to more complete mathematical approaches (Doak et al. 1998; McCann et al. 1998; Ives et al. 2000; Loreau 2000a; Ives and Hughes 2002), these results are qualitatively predicted from a model scenario conceptually similar to the insurance effect (fig. 9.1d).

As data on the effects of biodiversity on ecosystems become increasingly available and analyzed, these ideas will need to be refined. One of the more obvious issues is one of scale. In the empirical work done so far, the effects have been studied almost at the local scale. This means that the processes involved have almost entirely been due to population dynamics. However, in chapter 7 we saw that many of the large-scale patterns of community structure could be strongly affected by an additional process of species sorting within a regional context. The majority of experiments on the effects of diversity on both ecosystem function and stability have eliminated the species sorting process by randomly selecting species from a regional pool. We suspect that allowing species sorting to occur in both the theories and experiments will greatly enhance our understanding of these processes.

9.2. How Does Species Sorting Influence Ecosystem Properties?

Although large-scale ecosystem-level experiments, such as the manipulations of entire lakes (Schindler 1990; Carpenter et al. 2001) or large plots of land (J. H. Brown 1998; Sinclair et al. 2000), are generally considered to be more realistic than smaller-scale experimental studies, these experiments are often limited because they do not allow the process of species sorting. For example, several important studies have investigated the ecosystem-level responses of entire lakes to experimentally varied levels of nutrients (Schindler 1990), pH (Frost et al. 1999) or predators (Carpenter and Kitchell 1993; Carpenter et al. 2001). In these experiments, the species present in the lake at the start of the experiment are well suited for preexperimental conditions but may be less well suited to the new environment than other species that were not present in the preexperimental ecosystem. Of course, eventually we would predict that species favored in the changed system would eventually colonize, but this process could take a much longer time period than most experiments last.

As an example, Brown and colleagues set up large exclosures to eliminate kangaroo rats (*Dipodomis* sp.) from experimental plots in the deserts of the American southwest. For nearly twenty years, the ecosystem of these plots was strongly altered by the absence of this species (see J. H. Brown 1998 for a review). Eighteen years later, *Chaeotodipus baileyi*, a pocket mouse that was present at adjacent sites, invaded these experimental exclosures and compensated at an ecosystem level for the loss of the kangaroo rat (Ernest and Brown 2001). This experiment was designed to simulate what might happen if kangaroo rats went extinct. However, because the sorting process was not included in the experimental design, the initial results were much stronger than what would result if a real extinction occurred on a larger spatial scale (because some localities would already have the pocket mouse). Indeed, any experimental protocol that manipulates some aspect of the environment or biota of a particular ecosystem is probably similarly limited simply because the ecosystem is manipulated at much smaller and shorter scales than the process of species sorting acts. Much ecological theory suffers from the same limitation because it does not explicitly include the process of species sorting.

The Role of Species Sorting on Ecosystem Trophic Structure

Trophic structure is often considered to be an important indicator of ecosystem functioning because it describes the apportionment of energy and biomass among different trophic levels in the ecosystem (Elton 1927; Lindeman 1942; DeAngelis 1992; Pastor and Cohen 1997; de Mazancourt and Loreau 2000). One of the well-studied issues has been how trophic structure responds to enhanced supply of limiting nutrients to ecosystems. The problem has best been studied in terms of the relative apportionment of biomass in plants versus herbivores (although other trophic levels can also respond).

Models of food chains based on simple Lotka-Volterra predator-prey models (Oksanen et al. 1981; Oksanen and Oksanen 2000) predict that plants and herbivores (as well as other trophic levels) will respond differently to increasing levels of primary productivity and that this difference will depend strongly on the number of trophic levels in the ecosystem. In a simple food chain, the top trophic level is predicted to limit the trophic level below it, and increased productivity will pass directly into the top level. Thus, if there are two trophic levels (plants and herbivores), herbivores should increase while plants remain the same as productivity increases, whereas if there are three trophic levels (plants, herbivores, and predators), plants and predators should increase with herbivores remaining the same with increasing productivity (Oksanen et al. 1981). There is thus an asymmetric response between adjacent trophic levels with increasing primary productivity.

Empirical correlations between resource supply, plant biomass, and herbivore biomass typically show little evidence for such asymmetric responses (recent reviews of this evidence are found in Leibold et al. 1997; Chase, Leibold, Downing, and Shurin 2000), although interpretation of these results is controversial (Moen and Oksanen 1992; Oksanen and Oksanen 2000). A number of possible hypotheses, modifying the simple population dynamic models of stacked predator-prey models (i.e., food chains), suggest how deviations from the simple models may explain the lack of strong asymmetry in empirical findings:

(1) *Direct interference*, or nontrophic interactions among plants (e.g.,

allelopathy) and animals (e.g., territoriality), can lead to joint plant and herbivore responses (Gleeson 1994), although the number of trophic levels present still tends to cause asymmetry in the results.

(2) *Ratio dependence* suggests that a more appropriate way to model the functional response (the attack rates of consumers on their resources) is as a ratio-dependent function, rather than the typical prey-dependent function (Arditi and Ginzburg 1989). This produces joint responses that do not strongly depend on the number of trophic levels, but it creates other theoretical problems (Gleeson 1994; Abrams and Ginzburg 2000).

(3) *Inedible or resistant prey* could allow joint plant and herbivore response to enhanced resource supply (e.g., Leibold 1989; Hunter and Price 1992; Power 1992) because some of the benefits would be accrued by the resistant plants and some by the herbivores mediated through an enhanced turnover rate of edible plants (even if the edible plant biomass was constant).

(4) *Omnivory or intraguild predation* is the case of many species feeding on organisms from multiple trophic levels. Depending on how these organisms are assigned to trophic levels, models that include such interactions can produce joint correlations between plant and herbivore biomass (Holt and Polis 1997; McCann et al. 1998; Diehl and Feißel 2000).

(5) *Other food web–based configurations* exist. Abrams (1993) has modeled all possible configurations involving up to three trophic levels with up to two species per trophic level and found that the potential responses are indeed affected by these configurations. In many cases, it is possible to generate joint responses by plants and herbivores to enhanced resource supply.

(6) *Nonequilibrium dynamics,* where populations continuously cycle, show that the average long-term effects on trophic structure might not be the same as those expected on the basis of equilibrium assumptions (Abrams and Roth 1994). Under these conditions, joint average plant and herbivore responses to enhanced resource supply can occur.

All of the above hypotheses involve more complex interactions than the simple food chain models, and there is some empirical support for many of the hypotheses. Unfortunately, because many of the patterns predicted by the different models are similar, it is difficult to distinguish among the different hypotheses without more evidence. For example, Leibold et al. (1997) reviewed a variety of different types of evidence, combined with inductive reasoning, to discern which processes were most likely to be driving the patterns. First, they collated results from large-scale surveys of unmanipulated ecosystems and

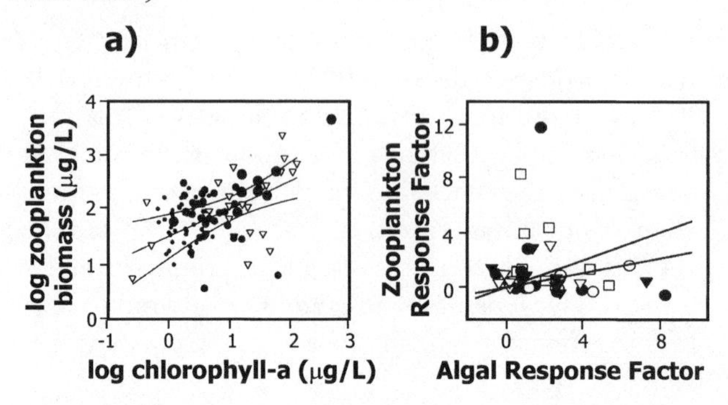

Fig. 9.2. Effects of increasing nutrients on algal and herbivore biomass in lake ecosystems. (a) Responses of algal biomass (measured as chlorophyll-a concentration) and herbivore biomass along a natural gradient of lakes with differing nutrient status. (b) Responses of algal and herbivore biomass to nutrient additions in lake ecosystems (experiments ranged from small short-term to long-term whole-lake experiments). The response factors represent the relative difference between control and nutrient addition treatments (i.e., an effect size). The solid lines represent the 95 percent confidence intervals from the survey data in part (a). Notice that the results from the experiments are much more asymmetrical than the natural surveys. Shapes and openness of symbols represent different types of experiments and are not relevant to this discussion. Both figures are redrawn from Leibold et al. 1997.

found little evidence for asymmetric responses in trophic level abundance along natural gradients; instead, plant and herbivore biomass were strongly correlated (Leibold et al. 1997) (fig. 9.2a). Second, they found that in experiments in lake ecosystems, the addition of nutrients rarely affected plants without also affecting herbivores. Furthermore, while most often plants and herbivores both responded positively to nutrient additions, these responses were variable; in about half the studies the plant response was much greater than that of the herbivores and in the other half the herbivore response was much stronger than that of the plants (Leibold et al. 1997) (fig. 9.2b). Whether plants or herbivores responded also did not seem to be strongly associated with either the number of trophic levels or with the occurrence of some taxa, such as members of the genus *Daphnia*, which are thought to be particularly important herbivores in these ecosystems. While ratio-dependent models are unlikely to explain these results, any or all of the other explanations listed above are reasonable to explain the observed experimental results in the absence of additional information.

So the question is not just *why don't the data from surveys of unmanipulated ecosystems and the theory match?* Instead, it becomes more complicated, and we ask *why don't survey data also match the data from field experiments?* We know that at least some of the variation in nutrient levels of the "unmanipulated ecosystems" is due to human activities (especially agriculture and sewage effluent). So in a way, those are unintended experiments quite similar to those purposefully conducted by scientists. An important difference is that the response to these unintended nutrient additions has occurred over much longer periods (decades or centuries) and at much larger spatial scales than the typical experimental manipulations.

We believe that the differences between the short-term responses of plants and herbivores from experimental manipulations of nutrients and the long-term responses of plants and herbivores to natural nutrient gradients may be due to factors related to the species sorting process. That is, the niche-based framework, and in particular the species sorting process, can help to resolve the discrepancy between experiments and surveys. Consider the module of keystone predation (Holt et al. 1994; Leibold 1996). In experimental manipulations, only the species currently present in the community are the ones that respond to the manipulation. If, for example, plants show a very large response, while herbivores show little or no response (such as several of the cases illustrated in fig. 9.2b), this could mean that there is a large population density of at least one plant taxon that is not well exploited by any of the resident herbivores. However, there might be an appropriate herbivore that does exist somewhere else but not in the experimental community. In the long term, after species sorting occurs and this herbivore colonizes the ecosystem, we would expect the plant biomass to decrease and the herbivore biomass to increase (see also Grover 1994).

While this hypothesis still needs to be tested, there are ancillary data that support this idea in both limnetic (Leibold et al. 1997) and terrestrial grassland ecosystems (Chase, Leibold, Downing, and Shurin 2000). The most important line of evidence is that there is quite a bit of compositional turnover in plants (and probably in herbivores) over gradients in resources supply and in the presence and absence of consumers. Additional evidence comes from studies conducted in artificial mesocosms. In some of these studies (Leibold and Wilbur 1992; Persson et al. 2001), the biota come not from a single source pond or lake but rather the biota are assembled by pooling species from a variety of sources (likely encompassing much of the regional species pool). These studies are exceptional in producing joint responses in plant and herbivore biomass that are approximately similar in relative strength to those observed in the surveys of unmanipulated systems.

Species Sorting and the Regulation of Available Inorganic Nutrient Pools

Nutrient cycling is a complex process that involves both biological and geological processes. One key parameter is the concentration of inorganic nutrients in an ecosystem, which strongly influences the rates of primary production and recycling. Any factor that affects the concentration of inorganic nutrients should alter many of the other key elements of the nutrient cycle. We can use the concept of species sorting within different niche-based modules to evaluate the primary structuring forces of natural ecosystems. As an example, we illustrate the role of species sorting on nutrient availability in limnetic ecosystems (Leibold 1997).

The regulation of nutrient availabilities due to competition among algae species for resources is well described in limnetic ecosystems. In a resource competition module, we assume that phosphorus (P) and nitrogen (N) act as essential resources (reviewed in Tilman 1982; Grover 1997). Thus, this module can be used to investigate how differences in relative ratios of nutrient supply jointly affect the composition of algae communities and conversely on the ability of algae to reduce the concentration of the available nutrients.

Given two limiting nutrients, we expect only one or two coexisting species of algae. In the absence of species sorting, we can model this situation along a range of resource supplies (fig. 9.3a). This simplistic model indicates that the N specialist will outcompete the P specialist when the relative supply of P is much greater than that of N, the P specialist will outcompete the N specialist when the relative supply of N is much greater than that of P, and the species will coexist when resource ratios are close to 1. We can use this model to predict the equilibrial concentration of nutrient availabilities that would be expected. When N:P is low, equilibrial nutrient concentrations will be fixed along the P specialist's ZNGI, whereas when N:P is high, the equilibrial nutrient concentrations will be fixed along the N specialist's ZNGI. Where they coexist, at intermediate N:P, both species' ZNGIs will determine the equilibrial nutrient concentrations. Thus, at equilibrium, we predict that N availability and P availability will be negatively correlated, with a very strong curve, due to the actions of the algal species under different N:P ratios (fig. 9.3b).

These predictions are slightly modified when we add the effects of species sorting (fig. 9.3c). In this case, there is compositional turnover across a gradient in N:P supply favoring N specialists when N:P is low, P specialists when N:P is high, and generalists at intermediate N:P. As above, the equilibrial nutrient availability will vary along the ZNGIs of the most favored species for any given nutrient supply. Thus, with species sorting among many species competing for two essential nutrients, we predict that there should be a reasonably smooth negative

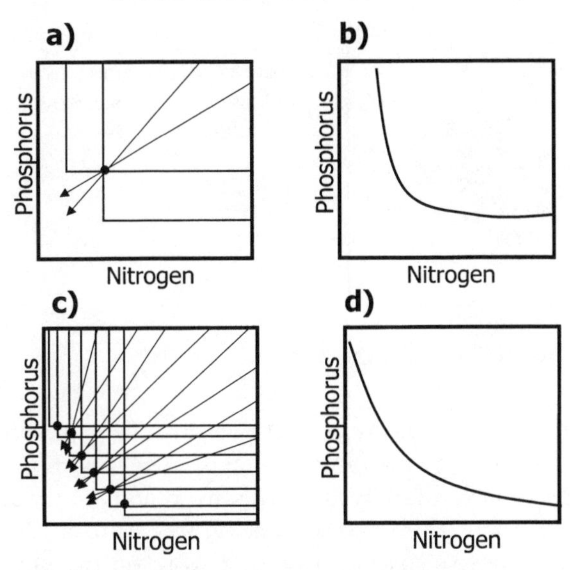

Fig. 9.3. Predicted equilibrial abundances of two primary limiting nutrients, nitrogen and phosphorus, in the case of many aquatic ecosystems based on the resource competition module. (a) Two species competing for the two essential resources (from Tilman 1982). (b) The predicted equilibrial availability of nitrogen and phosphorus in an ecosystem with these two species. (c) Several species competing for the same limiting nutrients; same scenario with a larger species pool and the opportunity for species sorting. (d) The predicted equilibrial availability of nitrogen and phosphorus in an ecosystem with all of these species. Note that while the relationship is smoother, both models predict a negative relationship between the availability of the two nutrients.

correlation between the availability of N and P when algal species are at equilibrium (fig. 9.3d).

In chemostat experiments, there is often the negative relationship, with a hard elbow, in resource availability that depends on pattern of coexistence and the N:P ratio, as predicted by the simple model (reviewed in Tilman 1982; Grover 1997). However, patterns from nutrient data of many lakes are the reverse of those predicted (fig. 9.4) (Leibold 1997). Instead of a negative correlation between N and P with either a hard curve (no sorting) or a smooth curve (with sorting), Leibold (1997) found a noisy but positive correlation. This mismatch between the theory, laboratory experiments, and observations of unmanipulated systems is similar to the mismatch that we saw in the previous section with trophic structure.

In contrast to the model of trophic structure, adding species sorting doesn't seem to help very much since negative correlations are still expected and these are inconsistent with the positive ones that were

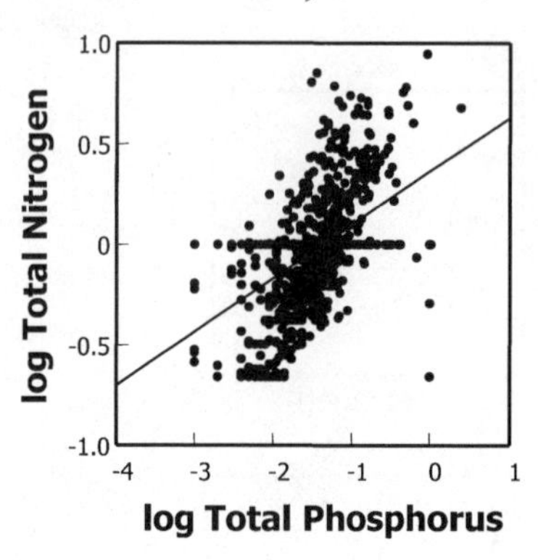

Fig. 9.4. The observed relationship between total nitrogen and total phosphorus availability from wide-ranging surveys of lake ecosystems; the slope of the relationship is significantly positive. Data from the surveys reviewed in Leibold 1997.

observed (fig. 9.4). There are at least two possible explanations (Leibold 1997). First, light competition could interact with competition for nutrients to alter these results. This hypothesis is thought to be especially important in terrestrial systems (see Tilman 1982, 1988; Huisman et al. 1999). Recent theoretical work on the interaction of competition for light with competition for nutrients (see Tilman 1988 for terrestrial plants and Huisman and Weissing 1995 for a very different model with similar fundamental results for phytoplankton) indicates that we should expect a negative correlation between light and nutrient availability. Empirical support for this prediction is found in the negative correlations seen between P availability (thought to often be the most strongly limiting nutrient in lakes) and light penetration (a measure of availability) in lakes (fig. 9.5).

Second, grazing by herbivores interacts with competition for nutrients. In the presence of grazers, absolute increases in nutrient levels (often correlated with resource ratios) will lead to increases in nutrient availability if grazing increases with nutrient supply. Our discussion of trophic structure indicates that this is quite likely, but this awaits experimental confirmation. Nevertheless, there is evidence that grazing can alter the correlations between the availabilities of different nutrients. In lakes that have undergone substantial acidification and consequent reductions in both biomass and diversity of zooplankton herbivores, we observe a negative correlation between P supply and N availability (fig. 9.6). Lakes in the same areas that have not undergone acidification show positive correlations instead (Leibold 1997). Of course, it is quite possible that other explanations for this difference between acidified and buffered lakes might be just as good. Our main point here is that an explicit niche-based perspective, combined with both deductive and inductive reasoning, can allow us to gain a much

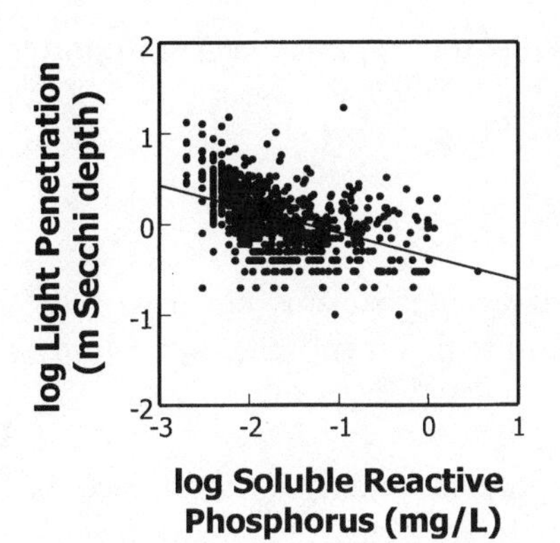

Fig. 9.5. A significantly negative relationship between light availability and soluble reactive phosphorus (a measure of phosphorus availability in lakes). From the surveys reviewed in Leibold 1997.

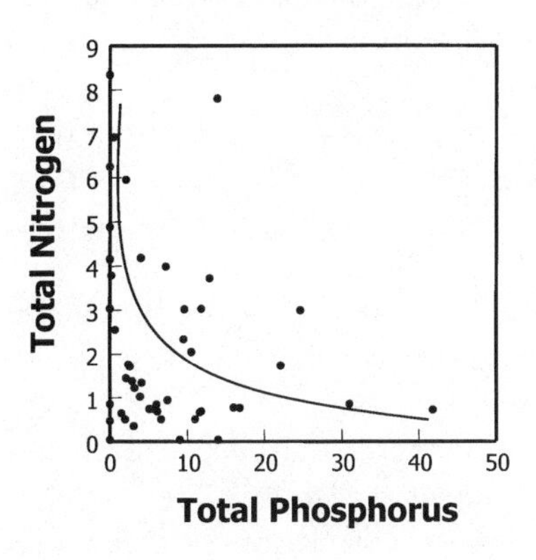

Fig. 9.6. A significantly negative relationship between total phosphorus and total nitrogen from a survey of lakes that were acidified and likely to have lost a considerable amount of their herbivore fauna. Data from the surveys reviewed in Leibold 1997.

better understanding of the role of species interactions in ecosystem-level processes.

9.3. Concluding Remarks

Ecosystem functioning involves much more than just species interactions. There is a huge array of biogeochemical processes and of landscape-level exchanges that are not easily addressed (or even necessary to address) using a niche-based approach. Our point is not that the niche concept can provide a framework for a complete understanding of what ecosystem functioning is and how it works. What we hope to have accomplished here is to show how the processes of species sorting at the community level can profoundly influence patterns of ecosystem functioning. To date, most studies on the influence of species diversity on ecosystem functioning have ignored the processes that determine community structure in the first place (Loreau 2000b; Loreau et al. 2001). The niche framework of species requirements and impacts provides a synthetic way to more lucidly traverse between community and ecosystems ecology (see also Tilman 1999; Chesson et al. 2002; Holt and Loreau 2002).

9.4. Summary

1) The niche framework illuminates the mechanisms underlying the effects of species diversity on ecosystem functioning. Specifically, we consider the niche complementarity effect, the insurance hypothesis, and the sampling effect.
2) Species sorting is an important process underlying the effects of species diversity, composition, and interspecific interactions on ecosystem attributes. Many experimental manipulations, which do not allow for the species sorting process, are of limited applicability to natural situations.
3) Experiments often find asymmetrical responses among different trophic levels to nutrient manipulations, whereas broad-scale observational studies find more symmetrical responses. We discuss data on trophic structure (herbivore and plant biomass) from limnetic ecosystems to illustrate this descrepancy. We suggest that differences in time scales, and whether or not species sorting has taken place, can account for these discrepancies.
4) We use the example of algae in limnetic ecosystems to evaluate the regulation of nitrogen and phosphorus availability. The resource competition module predicts a negative correlation between the availability of N and P at different levels of resource supply. The prediction is supported by laboratory experiments but not by large-scale observational studies.

CHAPTER TEN

THE EVOLUTIONARY NICHE

Species traits—be they something as important to the niche concept as resource consumption rates or something that might seem to have little ecological purpose, such as the number of tiny bristles on the thorax of fruit flies—are shaped by the processes of evolution. Throughout this book, our use of the niche concept critically depends on the traits of the species of interest. If the niche depends on traits and the traits are shaped by evolution, the obvious corollary is that the niche must depend on evolution. Hutchinson and MacArthur's contributions to ecological thought resulted, at least partly, from their focus on the evolutionary ecology of species' niches within communities (see chapter 1). Costs, benefits, and trade-offs, among other things, are just as important in an evolutionary context as they are in an ecological context.

The majority of the theory on the evolutionary ecology of niches only considers cases where a trait evolves adaptively. Evolutionary dynamics in any real case will be additionally influenced by less directed processes such as gene flow among adjacent habitats, mutation, genetic drift, and genetic linkage among traits. In this chapter, we deal less with nonadaptive cases than adaptive cases of niche evolution, but note that adding nonadaptive evolution to the framework can be accomplished in a similar way to adding ecological complexities (chapters 5 and 6).

One of the appeals of the niche framework as we have discussed it is that it connects the traits of species with their consequences for communities and ecosystems. Thus, we can, hopefully, explore not only how the evolutionary processes that influence species traits can manifest at the community level, but also how the community-level processes can influence the evolution of traits. When using traditional approaches to ecological theory (e.g., Lotka-Volterramodels), the connection between species traits and species interactions is not always obvious because the parameters (e.g., competition coefficients) are fairly abstract. While traditional models of evolutionary theory have clear parameters for species traits, they often ignore or "black box" their connection to the external environment and to associated species within the community. Explicit models of coevolution are an important exception (e.g., Thompson 1994; Rausher 2001). Interestingly, Laland et al. (1996, 1999) present a parallel approach, building from the evolutionary level, considering how species' impacts, which they term "niche construction," alter patterns of adaptation in simple population genetic models. Furthermore, when it has been explored, the predictions from a niche-based approach have much in common with those from related frameworks, such as adaptive dynamics, evolutionarily stable strategies (ESS), and optimization (Holt 1996; McPeek 1996; J. S. Brown 1998; Chase 1999b; Chase, Leibold, and Simms 2000).

In this chapter, we discuss how a niche-based approach can allow a direct exploration into a variety of factors that may influence the evolution of an individual's traits—such as exploring how changes in environmental conditions (e.g., resource supply), interacting species, or other factors can influence the traits of an organism. For this reason, the predictions that can be made from a mechanistic, niche-based framework can provide a significantly more powerful framework than those that do not explicitly consider the link between evolution, the biotic and abiotic environment, and populations, communities, and ecosystems.

10.1. Evolutionary Species Sorting

In chapter 7, we introduced the concept of species sorting at the community level. There, species' traits were fixed, and community-level sorting culled ecological duds out of the regional species pool. This sorting in turn determined the distribution and composition of species along environmental gradients. If we extend the niche concept to consider sorting at the population level, we can examine the process of phenotype sorting and the maintenance of phenotypic composition and diversity. Indeed, recent theoretical analyses have shown that the maintenance of species diversity at the community level has many similarities with the maintenance of phenotypic diversity at the population level (Hastings and Gavrilets 1999; Amarasekare 2000).

Remember that in our discussions of the interactions among different sorts of species, we are really talking about the relative fitness (as determined by ZNGIs and impacts) of species with different sorts of traits. Thus, we could just as easily be talking about the costs and benefits of phenotypic differences within a single species to discover the expected fitness advantages of those phenotypes in different environments (see also Tilman 1982; Grover 1997; Leibold 1998; Chase 1999b; Chase, Leibold, and Simms 2000). This phenotype sorting is analogous to concepts used in theories of trait evolution, such as an ESS (e.g., Maynard Smith 1976; Vincent and Brown 1988; Brown and Vincent 1987, 1992; Abrams 2001b). The ESS concept considers the array of possible phenotypes in a population, as well as the conditions in which a single or multiple phenotypes are favored and uninvasible by other strategies.

To illustrate the process of phenotype sorting, consider any module of interactions among phenotypes instead of species (fig. 10.1). We can determine which phenotypes will be expected to survive under different environmental conditions. For example, in fig. 10.1, the environmental conditions will determine which phenotypes are favored (ESSs), whether more than one phenotype can coexist locally or regionally (conditions for coexistence), which phenotypes can coexist (invasible or uninvasible equilibria), and whether any phenotypes are duds that will be eliminated from the population. Of course, this situation only represents the simplest ESS, and due to a variety of evolutionary constraints (e.g., drift, spatial structure, recombination), many complexities can be included to more fully explore the range of ESSs in natural systems. Nevertheless, the simple ESS approach allows a starting point for addressing a number of important issues in evolutionary ecology and in some ways can provide predictions similar to those of more complex evolutionary approaches such as quantitative genetics and adaptive dynamics (Taper and Case 1985; Charlesworth 1990; Vincent et al. 1993; Abrams 2001b).

10.2. Evolution and Species Interactions

If adaptation worked perfectly, we might expect that a single phenotype could adapt to the environmental conditions at precisely the trait combinations where it would be the most fit phenotype and drive all others extinct. Although such perfect adaptation would reduce local diversity to one phenotype, regional diversity could be maintained or even enhanced over a heterogeneous gradient (Thompson 1994). Perfect local adaptation does not seem to generally be the case, perhaps because it is extremely unlikely that the environment will ever be constant enough for a single phenotype to become unilaterally optimized to a given situation or perhaps because other genetic mechanisms (constraints, gene flow, sexual selection, etc.) prevent it. Variance through space

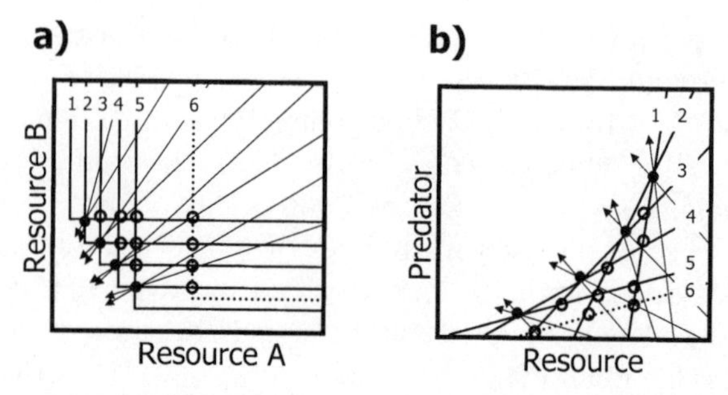

Fig. 10.1. Two examples of the evolutionary sorting process. In each case, ZNGIs are labeled for five different phenotypic strategies that show trade-offs in their traits (1–5) and an evolutionary dud (6), which is not expected to persist. (a) Competition for two essential resources. (b) Competition for shared resources and shared predators.

and time allows phenotypes to persist in environments to which they are not perfectly adapted, just as it did when we were discussing the coexistence of species in variable environments (chapter 6). In addition, even if a species completely adapts to the local environmental conditions, several evolutionary forces can act alone and in concert to cause divergence among the species traits, including assortative mating, mutation, and genetic drift (Dieckmann and Doebeli 1999; Case and Taper 2000; Schluter 2000, 2001).

Although initial links between community ecology and evolutionary niche concepts were very tight (Hutchinson 1959, 1965; MacArthur and Levins 1967; Levins 1968; MacArthur 1972), the two fields have slowly drifted apart. Today, most community ecological investigations take the phenotype of a species as a given and explore the interactions among species, whereas studies of the evolution of phenotypes typically black-box the ecological forces that drive selection on traits (but see Thompson 1994; McPeek and Miller 1996; Miller and Travis 1996; Schluter 2000; Day et al. 2002; Neuhauser et al. 2003).

10.3. Examples of the Utility of the Evolutionary Niche Concept in the Keystone Predator Module

In the remainder of the chapter, we discuss three evolutionary topics using the keystone predator module: (1) evolutionary character displacement among potentially coexisting species, (2) the evolution of defense versus tolerance to predation or herbivory, and (3) the evolution of phenotypic plasticity in response to predation. These ideas are preliminary, and we intend this section primarily to serve as a springboard for further investigations on the linkages between the ecological and evolutionary niche concepts. Furthermore, other modules of species interactions, such as resource competition or stresses, as well as questions regarding other evolutionary phenomena, could give equally interesting insights.

But first, we take a brief detour into the mathematics behind the graphical models that we have emphasized. We do this to make the parameters of the model explicit, and to relate these parameters to specific traits that can evolve. The keystone predator food web consists of a basal resource (R), where $c_j[S - R]$ describes the turnover rate of the resource in the supply pool (S) that is not tied up in R, any number of intermediate phenotypes of a species, which we will call consumer-prey (N_i), where the subscript i indicates the identity of a particular phenotype, and a predator species (P). We use a simple representation of the traits among these phenotypes to illustrate the key concepts and refer the readers to more rigorous mathematical treatments (Holt et al. 1994; Leibold 1996, 1998; Grover and Holt 1998; Abrams 1999; Chase, Leibold, and Simms 2000). Thus, we define the parameter f_i to describe the per capita attack rate of the consumer-prey phenotype i on the resource, m_i to describe the per capita attack rate of the predator on consumer-prey phenotype i, a_i to describe the per capita conversion rate of consumed resource into new consumer-prey phenotype i, c_i to describe the per capita conversion rate of consumed consumer-prey phenotype into new predators, and d_i and d_P to describe the density-independent loss (death) rates of consumer-prey phenotype i and predators. Given these definitions of some of the phenotypic traits, we can write a dynamical set of differential equations of the form

$$dP/dt = P(\Sigma(m_i c_i N_i) - d_P)$$
$$dN_i/dt = N_i(f_i a_i R - m_i P - d_i)$$
$$dR/dt = c[S - R] - \Sigma(f_i N_i).$$

The resource and predator densities that satisfy each phenotype's requirements and the species per capita impacts can be described by:

$$ZNGI_{(i)}: P^* = (f_i a_i R^* - d_i)/m_i$$
$$I_{(i)}: [f_i R,\ m_i c_i P].$$

The slope of the ZNGI, its intercept on the R axis, and the slope of the impact vector are:

$$\text{slope of } ZNGI_{(i)} = f_i a_i / m_i$$
$$ZNGI_{(i)} \text{ intercept on } R \text{ axis } (R^*) = m_i/(f_i a_i)$$
$$\text{slope of the impact vector} = m_i c_i P/(f_i R).$$

We now use these equations to illuminate discussion on three problems: character displacement, resistance or tolerance to herbivory, and phenotypic plasticity.

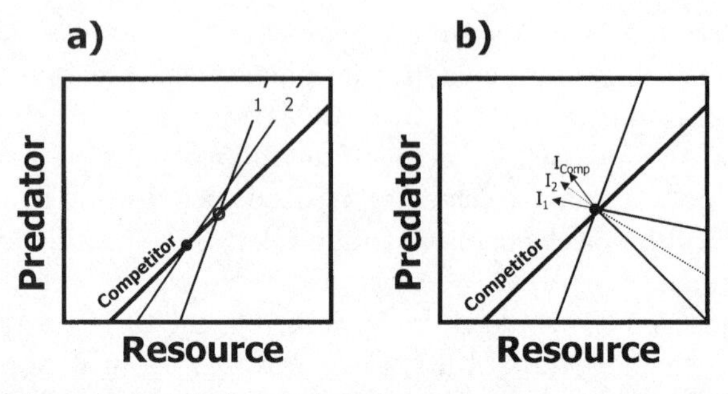

Fig. 10.2. Evolutionary character displacement in the keystone predator module. A competitor species is competed with two phenotypes of a focal species. (a) The phenotype whose ZNGI is most similar to that of the competitor species (uninvasible equilibrium denoted by a closed circle) will outcompete the phenotype whose ZNGI is most different (invasible equilibrium denoted by an open circle). (b) The phenotype whose impact is most different from that of the competitor species will exist over a wider range of environmental supplies, and likely drive the phenotype whose impact is most similar to the competitor species (dashed line) extinct.

Character Displacement

Character displacement (the name first appears in Brown and Wilson 1956) refers to a hypothesized relationship among coexisting species that evolve differences in their traits (characters) as a result of their interactions with each other; it has played a prominent role in classical and contemporary studies of niche evolution and adaptive species radiations (Lack 1947; Grant 1986; Schluter 1996, 2000, 2001; Losos et al. 1998; Case and Taper 2000; Losos and Schluter 2000). Using phenotype sorting, we predict that in a population with phenotypic diversity, the phenotype with the ZNGI most similar to that of the average phenotype of the competitor species will be favored (fig. 10.2a). However, other processes, such as environmental heterogeneity, might select for more divergent phenotypes (see also May and MacArthur 1972). In addition, the competing species are likely to coexist over a broader range of environmental conditions when their impacts are most divergent (fig. 10.2b). Thus, at the population level, the phenotype with the most similar ZNGIs but most divergent impacts from the average phenotype of the competitor species will be favored.

We can explore this using the mathematics discussed above, by examining the simple case where there are two phenotypes of consumer-prey (labeled with subscripts 1 and 2). If selection acts to sort out phenotypes that are most different in requirements, this will minimize the difference between the slopes of two consumer-preys' ZNGIs, and thus $f_2a_2m_1/f_1a_1m_2$ should be minimized (but greater than 1). Similarly, if selection acts to sort out phenotypes that are most similar in impacts, this will maximize the differences in the slopes of the impact vectors at any ZNGI intersection, and thus $m_2c_2f_1/m_1c_1f_2$ should be large. Some algebraic manipulation shows that the differences in ZNGIs will be minimized while differences in impacts is maximized will occur when the ratio a_1c_2/a_2c_1 is minimized (Leibold 1998). That is, selection should act so that the consumer-prey phenotype that has greatest conversion efficiency for the basal resource (a) also provides the best food for the predator (c).

Resistance versus Tolerance to Predation or Herbivory

Plant-herbivore studies have focused on two ways in which plants can evolutionarily respond to herbivory. They can resist herbivory through physical or chemical means, or they can tolerate herbivory through compensation (Strauss and Agrawal 1999; Stowe et al. 2000; Juenger and Lennartsson 2000). Similar mechanisms can also occur with predators and prey. In the keystone predator module, two consumer-prey phenotypes can potentially coexist when their ZNGIs intersect. Because the slope of the ZNGI of a phenotype is given by f_ia_i/m_i, a species can have a steeper ZNGI by (1) decreasing the predator attack rate (m) through a resistance mechanism or (2) having a higher capacity for tolerance at the population or individual level (in the case of plants) by increasing the product of the resource consumption (f) and conversion (a) rates. When there is a cost to either tolerance or resistance, the better-defended phenotypes (whose ZNGIs have steeper slopes) will be the poorer competitors (with higher R^*). Since $R^* = m_i/(f_ia_i)$ (where P is set to zero in the above equations), this cost (higher R^*), will occur with either an increase in the density independent loss rate (d), or a decrease in the product of the resource attack (f) or conversion (a) rates (Chase, Leibold, and Simms 2000).

For example, if we assume there are two phenotypes in a population, one that is more tolerant to predation and the other that is more resistant, we can explore the environmental conditions that will favor dominance by one or the other phenotype, or their coexistence. First, if the costs are lower (lower R^*) and the benefits are greater (steeper ZNGI) for either tolerance or resistance traits, then we would expect that only one trait would be favored under all environmental conditions (fig. 10.3a, b). Second, two scenarios can occur if there is a trade-off among tolerance and resistance traits, as is often expected among plants (Stowe et al. 2000). First, the cost of tolerance may be less than the cost of resistance, while the benefits favor the resistant phenotype. Here, the resistant phenotype is better defended from predators and also provides worse food for the predators (its impact is shallower than that of the

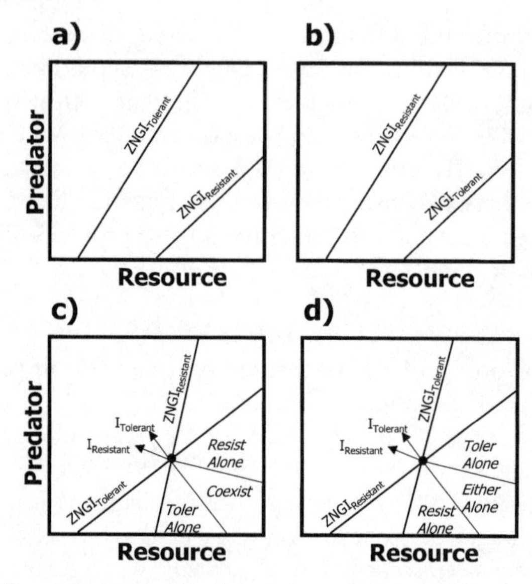

Fig. 10.3. Interactions among two prey phenotypes that compete for a common resource and are consumed by a common predator; one is tolerant to predation ($ZNGI_{Tolerant}$) and the other is resistant to predation ($ZNGI_{Resistant}$). (a) The costs and benefits favor the tolerant phenotype. (b) The costs and benefits favor the resistant phenotype. (c) The costs and benefits are more balanced, and the outcome of interactions between the two phenotypes depends on the resource supply. (d) Same as (c), except there is no coexistence, since the tolerant phenotype also provides better food for the predators.

other phenotype), and we predict that the two phenotypes can coexist under intermediate resource supply, while the tolerant phenotype is expected to exist alone at low resource supply, and the resistant phenotype is expected to exist alone at high resource supply (fig. 10.3c). Second, the cost of tolerance may be greater than the cost of resistance, but the benefits may favor the tolerant phenotype. Here, the tolerant phenotype is better defended against predators but also provides better food for them, such that at intermediate productivity the two phenotypes will not coexist locally, but instead alternative stable equilibria will occur. In addition, at low productivity the resistant phenotype will exist alone, and at high productivity the tolerant phenotype will exist alone (fig. 10.3d). Interestingly, analogous results were obtained by Tiffin (2000) in a genetic model of the evolution of tolerance and resistance traits among plant species.

Phenotypic Plasticity

Phenotypic plasticity occurs when a species can attain more than one phenotype depending upon the environmental situation in which

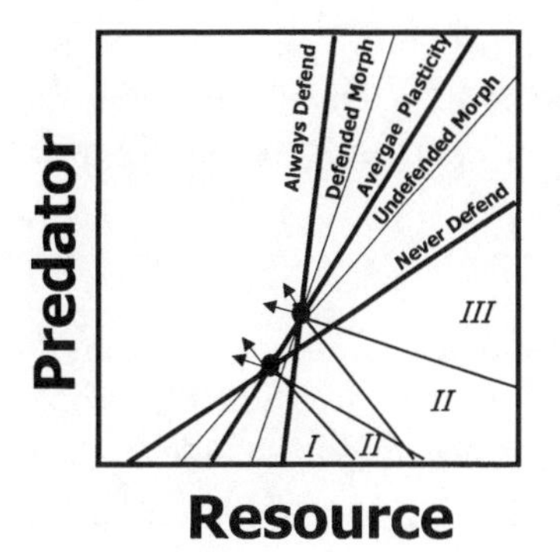

Fig. 10.4. Depiction of three strategies for defense against predators in the keystone predator module. Each strategy has assumed costs and benefits. I. The "never defend" strategy is favored. II. The plasticity strategy will coexist with one of the other strategies. III. The "always defend" strategy is favored.

it finds itself (see chapter 5). Recently, evolutionary biologists have become increasingly interested in studying the evolution and maintenance of plasticity. Of particular interest is how both animals and plants can phenotypically respond to the presence of an enemy (i.e., herbivore or predator) by changing its morphology, physiology, or life history. However, recent evidence has shown that plasticity can often be costly to an organism (e.g., Schlichting and Pigliucci 1998; Agrawal et al. 1999; Van Buskirk 2000). Furthermore, while some species in some environments maintain strong phenotypic plasticity, other species in other environments are more canalized and do not express the same degree of plasticity (reviewed in Agrawal 2001).

Using the keystone predator module, we can explore some circumstances whereby phenotypic plasticity may or may not be expected to persist as a viable phenotypic trait. Assuming both defense and plasticity are costly, but that pure defense is more so, we consider three strategies: (1) never defend, but be a uniformly better competitor, (2) always defend, but be a uniformly bad competitor, and (3) maintain phenotypic plasticity, with one morph that is relatively more defended and one that is relatively less defended but a better competitor (fig. 10.4). If resources are generally rare, such as in low-productivity areas, we expect only strategy 1 to be favored. Alternatively, if resources are abundant (but so are predators), we expect that strategy 2 will be favored. However, at intermediate levels of resource supply, we expect that strategy 3 can coexist stably with either strategy 1 or 2 locally when defense is through resistance or unstably when defense is through tolerance. In addition, when there is considerable temporal or spatial variation in the abundance of food or predators and the benefits of being plastic outweigh the costs, plastic strategies might be expected to outcompete the fixed strategies. Note, however, that the costs and benefits of plasticity and the relative position of the plasticity ZNGI may be altered by the reliability of the cues that prey utilize to decide which strategy to use (e.g., Lively et al. 2000).

10.4. Some Unanswered Questions

There are numerous other ideas on topics that relate the evolution of species traits to their ecological context. For example, some questions that we could ask with a niche-based approach might include:

1. How might reciprocal selection among species (coevolution) differ from asymmetric selection, where only one species is responding to the biotic and abiotic environment?

Our discussion of character displacement and species interactions does not consider whether a species is evolving independently, or whether there are reciprocal, coevolutionary interactions. The predicted results might differ depending on whether selection is acting on one or all species involved in an interaction.

2. How does the form of the ecological interaction module (e.g., resource competition, keystone predation, mutualism) influence the selective pressures and evolutionary traits?

Above, we used the keystone predator module of species interactions to discuss a few issues of interest to evolutionary ecologists. Some patterns of evolutionary dynamics are expected to be similar in all modules, such as the phenomenon of convergent requirements and divergent impacts among coexisting species (Leibold 1998). However, other patterns, such as the conditions in which phenotypic plasticity is expected to evolve, will depend critically on the types of limiting factors and the nature in which species are interacting.

3. How might the geographic (spatial) and temporal structure of species and their interactions shape the evolution of their niches?

As we discussed in chapter 7, spatial and temporal heterogeneities can play a strong role in the coexistence of species with different sorts of traits. In the coevolutionary literature, such spatial and temporal structure also plays a strong role (e.g., Wright 1931; Thompson 1994). Using the niche concept, we might discuss the constraints on adaptive niche evolution that result from geographical and temporal variation. For example, Holt and colleagues have shown how dispersal among habitat patches that are sources for a population can increase the probability of adaptation in habitats that are sinks for that population (Holt and Gaines 1992; Holt 1996; Holt and Gomulkiewicz 1997; Gomulkiewicz et al. 1999). Thus, connections among habitat types can strongly influence how species evolve to utilize those habitats, as well as how they interact and coevolve with other species (Gomulkiewicz et al. 2000).

4. How might the niche framework alter our predictions about evolutionary diversification of species on different trophic levels?

The traditional notion of evolutionary diversification suggests that terrestrial plants evolved diversity first, which allowed for the subsequent radiation of herbivores. However, this is a noninteractive bottom-up view of the diversification process. If, on the other hand, species interactions play a strong role in evolutionary diversification (see, e.g., Losos et al. 1998; Losos and Schluter 2000; Schluter 2000, 2001), as our discussion here assumes, we would predict that those radiations might more likely have taken place in a highly interactive coevolutionary process. This is because species become limiting factors for other species and thus parts of their niches. For example, a specialist insect herbivore can impede the growth of the plant on which it specializes and create opportunity for invasion by other plant species in much the same manner as occurs in Grover's (1994) community assembly scenario. This prediction, albeit highly speculative, is probably a testable one that can be compared with the null, noninteractive view of evolutionary diversification over evolutionary time scales.

5. How can we link ecological and microevolutionary concepts with the long time scales involved in macroevolutionary diversifications?

The niche concept has played a prominent role in the study of evolution over macroevolutionary time periods (Simpson 1953; Vermeij 1987; Jackson et al. 1996; Jablonski and Sepkoski 1996; Jackson and Johnson 2000; Alroy et al. 2001). However, in fossil communities, there is no guarantee that the organisms involved actually coexisted at a single point in time (due to temporal averaging) or space (due to spatial mixing). Consequently, paleontologists have always been very skeptical of conventional niche theory that is so strongly based on assumptions about local coexistence and focused on the relative impacts of species (chapter 3). However, in the niche-based approach we have discussed here, aspects of ecology that relate to relative impacts of species are much less important at larger scales than they are at local scales (see chapters 8 and 9) and aspects of ecology related to relative responses are much more important. We believe that a number of intriguing evolutionary hypotheses might benefit from this approach including the concepts of faunal mixing (Vermeij 1991), evolutionary escalation (Van Valen 1973; Vermeij 1987), speciation and evolutionary radiations (Alroy et al. 2001; Schluter 2000, 2001; Godfray and Lawton 2001), evolution related to habitat gradients (Jablonski et al. 1983), and the shifting dynamics of communities in the fossil record (Davis 1981, 1986; Jackson et al. 1996; Jackson and Johnson 2000).

10.5. Concluding Remarks

The niche concept has much to offer studies of evolutionary biology. Similarly, there is much that ecologists can learn from careful attention to evolutionary processes, and we feel that the most important step is to have a conceptual language that can bring more evolutionary thinking into the complexities of species interactions in communities and ecosystems. Because the niche concept, in an ecological context, has many similarities to the niche concept in an evolutionary context, and both are reliant upon fitness, we feel that it is one such currency.

Interestingly, a series of theoretical studies by Laland, Odling-Smee, and Feldman (Laland et al. 1996, 1999; Odling-Smee et al. 1996) have utilized standard population genetic models and added complexity that they term "niche construction." Their niche construction is essentially the effect of a species on its environment, and corresponds very nicely with our definition of the impact component of a species' niche. Thus far, their models are somewhat simplistic in that they only consider the evolution of traits at two loci with two alleles each, and they do not explicitly allow for coevolutionary processes among species that have mutual impacts; nevertheless, we expect that models of this form will be highly complementary to our framework.

Currently, a majority of evolutionary ecology experiments aimed at exploring the role of species interactions in trait evolution are performed much as they would be when exploring the role of abiotic factors. This generally means holding the density (or consumptive effects) of the interacting species constant and measuring the response of the target species. However, recent theoretical models show that the expected evolutionary response of a target species might be very different if the interacting species is allowed to numerically respond to the traits of the prey than if it cannot respond (Tiffin 2000; Chase, Leibold, and Simms 2000; Day et al. 2000). For example, in the plant resistance/tolerance model, the favored phenotype results from an interaction between competition for resources and apparent competition, which results from herbivores responding numerically to plant traits. Thus, experiments that do not allow herbivores to numerically respond to plant traits may not give a true indication of how plants evolutionarily respond to herbivory. In fact, this is likely to be true for a wide variety of experimental studies on the evolutionary ecology of predator-prey interactions, which often do not allow the predator or the prey to respond numerically to the interaction (see also Chase and Knight 2003). A solution to the current mismatch between experimental and theoretical approaches will require the sort of creativity and pluralism outlined in chapter 4, integrating theoretical modeling, observations, and short- and long-term experimentation.

10.6. Summary

1) The concept of phenotypic sorting is identical to that of species sorting but considers the distribution and coexistence of phenotypes along environmental gradients. Determining the equilibrium phenotype(s) expected in a locality is analogous to an evolutionarily stable strategy (ESS) modeling approach.
2) By exploring the costs and benefits of traits along environmental gradients and with interacting species, we provide an explicit link between the evolution of species' traits and the ecological context of the selective environment.
3) The keystone predator module serves as an example to illustrate the utility of the niche approach for the evolution of phenotypes.
    a) In sympatry, selection should act to minimize the difference between two species' ZNGIs but maximize the difference between their impacts.
    b) The evolution of two different phenotypes, resistant and tolerant, to predation or herbivory will depend on the relative costs and benefits of each trait. When resistance is more costly (but more beneficial), then both phenotypes can coexist in the same locality, whereas when tolerance is more costly (but more beneficial), the phenotypes cannot coexist locally but can regionally.
    c) When plasticity in the amount of investment to predator defense is costly to a phenotype, its expression will depend on the level of resource supply. When resources are rare or common, a species should never be plastic, and be undefended or defended, respectively. However, at intermediate resource supply, the species should exhibit plasticity to take advantage of variable resource and predator conditions.
4) Our definition of niche might lend insight into a variety of other questions in evolutionary ecology. This is because it explicitly connects ecological and evolutionary scales by having fitness as its common currency.

CHAPTER ELEVEN
CONCLUSIONS

In chapter 1, we discussed how the use of the word *niche* has declined in ecology. Clearly, ecology has advanced beyond the works of Hutchinson, MacArthur, and others who used the niche concept as a framework for ecological study, and many of their ideas and results have since been found to have shortcomings (see, e.g., Gotelli and Graves 1996). However, many ideas and concepts that were born in that era still have relevance to modern ecological study. In this book, we have put forward a revised view of the niche concept that we think can serve to reevaluate and reinterpret a number of these ideas.

Perhaps we can best illustrate the historical agnosticism of much of current ecology with an example and comparison with its sister science, evolutionary biology. The vast majority of current evolutionary biology is still strongly influenced by the "modern synthesis" of evolution, which occurred most notably due to Fisher, Wright, Simpson, Dobzhansky, and others during the middle part of the last century (over fifty years ago) (Mayr and Provine 1998). If there is such a synthesis in ecology, its existence is not apparent. Indeed, the contrast between ecology and evolutionary biology in their reliance on historical concepts can be easily seen in any recent journal. To illustrate this, we randomly chose three issues of the journal *Ecology* and the journal *Evolution* from the year 2000. We found that in *Evolution*, over 30 percent of the references were pre-1980, and nearly 10 percent were pre-1960. In contrast, fewer than 10 percent of the citations were pre-1980 and fewer than 1 percent were pre-1960 in *Ecology*.

Ecology is a field that is rapidly coming into its own, even if perhaps not for the first time. The general public is increasingly interested in environmental topics, including the patterns and loss of species diversity and the moral, aesthetic, and economic value of species diversity and ecosystem function. In this exciting and diverse intellectual bazaar, there are going to be an increasing number of perspectives and approaches that will influence where the field goes (Thompson et al. 2001). Obviously, we feel that the niche concept is an important ingredient in this brew. This is because it provides a currency that can incorporate and synthesize many seemingly disparate ideas ranging from the individual to the ecosystems level.

11.1. What Does Our View of the Niche Concept Do for Ecology?

We have taken a very different approach to the niche from the one that is most often taken in textbooks. We feel that this approach has several important advantages. Specifically, our niche framework provides new insights and interpretations about ecological patterns and processes in ecology, fixes some important misconceptions in ecological thinking, synthesizes the insights made by many people into a common language, and allows for flexibility in bridging questions from the individual to the ecosystem level.

First, our approach provides a much closer link between the niche concept and mechanistic models of population regulation and species interactions than conventional niche theory. Rather than use vague concepts such as "niche overlap" and "niche breadth" that involve a number of assumptions about the context in which interactions occur, we focus on measurable aspects of the biology of organisms such as growth rates, consumption rates, and death rates. The links to these mechanistic models also illustrate that a much broader array of mechanisms can be investigated than are possible with the conventional approach, including the roles of predation, mutualism, abiotic stresses, and bioengineering. While these mechanisms involve a diverse array of processes, our approach to the niche illustrates how their consequences can be viewed in a single framework because of their common effects on either the responses of organisms to their environment, the impacts of organisms on their environment, or both.

Second, we feel that our approach is flexible enough to deal with a wide array of issues that were difficult to address with previous niche concepts. These include complex patterns of behavior, physiology, and population structure as well as the ability to capture some of the more important ways that fluctuating environments might regulate population dynamics and species interactions. Perhaps even more importantly, our approach can help elucidate the role of species interactions at spatial scales larger than the local community level.

Unlike previous niche concepts, our approach explicitly recognizes the inherent duality of the niche, contrasting the ways that organisms respond to their environment (i.e., their requirements) with the reciprocal ways that they impact their environment. Nevertheless, our approach has very strong roots in the previous literature. Much of the ideas and conceptual models that we have discussed in this book are unashamedly borrowed from others. Our intent has not been to reinvent niche-based theory, but rather to take a variety of approaches and questions and weave them all into a single synthetic framework. Some of the most important insights that are gained include:

1) The distinct roles that requirement components and impact components of the niche play in how communities assemble. Whether a species can establish in a community is more related to its requirements, whereas the stability of that community is more related to its impacts.
2) The application of the niche concept to thinking about community structure at both local and regional spatial scales. At the local scale both requirement and impact components determine which species can exist, whereas at the regional scale the requirement component is much more important than the impact component. This also leads to the prediction that patterns of biodiversity might change with spatial scale under many circumstances.
3) The important ways that context dependence alters species interactions. The strength of interspecific interactions depends on both the environmental context (related to the supply points), the composition of the community, and the other types of processes that are simultaneously occurring (e.g., food competition vs. predation vs. stress).
4) The importance of species sorting for linking attributes of larger scale biotas to the dynamics of local communities and ecosystems. Especially revealing is the importance of the trade-offs that are present in the regional species pool (rather than simply in the local community) for understanding how communities and ecosystems might change in time and space.

11.2. Limitations to the Niche Framework

The advantages that we see in our synthetic niche-based framework do not come for free. We have tried to discuss several limitations up front during our descriptions of the models and ideas, rather than hide them. However, the limitations of our approach are not a consequence of the definition we use for the niche or of the basic framework, but rather are a consequence of the tricks that must be developed in order to develop intuitive predictions with broad implications. For example, we have limited our primary focus to conditions of zero net population growth and linear species interactions. We did this in order to gain as much as possible from the graphical depiction of these models. While these might be reasonable approximations at some scales, they cannot capture all of the complexities that one might want.

We certainly know that both theoreticians and empiricists often get frustrated with this generalized approach. As it becomes necessary to consider the complexities, theory will require a more rigorous mathematical framework and empiricism will need more detailed studies. In some ways, the graphical models that we focus on can be thought of as generating ideas rather than explicit proof. If those ideas turn out to be interesting or important, then one could turn to the details and more rigor. However, the complex mathematical models will take much the same form as our models (albeit with increased numbers of terms), since they are both based on consumer-resource dynamical systems (see chapter 2). Similarly, many of the complexities of natural systems do not negate our approach but instead add a richer array of possible outcomes (see chapters 5 and 6). And thus, our graphical approach to the question can serve as a benchmark from which one can evaluate the consequences of these complexities.

Throughout we have primarily focused on species interactions in the context of coexistence. However, ecologists are often interested in the relative abundance of species (i.e., how many species are common or rare), which we have mostly ignored. In fact, this is one of the great appeals of Hubbell's (2001) approach because it predicts not only species coexistence but also relative species abundances. Of course, the basic models that we have presented are intimately linked with relative species abundances, but our graphical depiction of them makes this somewhat difficult to intuit. For example, when resource supplies are interior to the range of coexistence between a pair of species, the species will coexist, but their relative abundances will change throughout the range of coexistence, which depends on the relative availability of the factors. Nevertheless, we foresee that combining the patterns of interaction and coexistence that we have primarily focused on here with a knowledge of relative species abundances should provide considerable grist for future studies.

11.3. Niche versus Neutral Theory

Robert MacArthur was a pioneer in both the niche theory (MacArthur 1958, 1965, 1969; MacArthur and Levins 1967) and the theory of biogeographical ecology (MacArthur and Wilson 1967; MacArthur

1972). However, he rarely made the link between these fields and in his studies at large biogeographic scales typically assumed that there were no differences among species in their niches; they simply followed a probability distribution of extinction and colonization processes. The conceptual chasm between the two paradigms has been termed "MacArthur's paradox" (Schoener 1989; Loreau and Mouquet 1999).

Recently, Hubbell (2001) has appropriated many of the ideas from MacArthur and Wilson's (1967) theory of island biogeography to develop a theory that incorporates immigration and extinction processes, which he terms "ecological drift," as well as speciation but contains no assumptions of niche differences among constituent species within the metacommunity (see also Bell 2000, 2001). Hubbell's species are thus neutral with respect to each other, and he argues that the process of ecological drift can be an extremely useful tool for understanding some of the most fundamental questions of ecology such as species diversity and abundance. In many ways, Hubbell's approach is the exact inverse of ours. Hubbell certainly does not discount the importance of the niche concept in his discussion, although his models do not require any niche differences to generate predictions. Hubbell's stated goal was to see how far he could go with his oversimplified assumption that species traits were neutral. One of the great appeals of Hubbell's approach is that it appears to explain a number of important patterns on both the local diversity and abundance of species. In particular, predictions from Hubbell's model are consistent with *some* of the patterns seen in the great diversity of tree species and distributions of tree species abundance in many tropical forests and other systems (Hubbell 1997, 2001; Bell 2000, 2001). However, the robustness of Hubbell's predictions is limited to certain model parameters that await empirical verification (Zhang and Lin 1997), and it is unclear exactly how different the predictions of many natural patterns would be between neutral and niche models (Brokaw and Busing 2000; Chave et al. 2002). In addition, other studies in tropical rain forests, the very system that Hubbell's model was intended to explain, are often not consistent with the neutral model and suggest a stronger role for niche-based processes (e.g., Terborgh et al. 1996; Yu et al. 1998; Pitman et al. 2001; Condit et al. 2002).

Our presentation has especially focused on the ways that the niche concept allows us to describe and evaluate the consequences of trade-offs among species in the ways they respond to and affect aspects of their environment. Further, we assume that such trade-offs are important in generating variability among communities and in regulating the relative abundances and distributions of species. In essence, our goal has been to elucidate just how far we can go with a reasonably simple framework based on niche requirements and impacts.

The differences between our approach and Hubbell's (2001) may seem irreconcilable at the outset. However, we hope for a different response. We echo Hubbell's recognition that the niche and the neutral models are in reality two ends of a conceptual continuum with the truth most likely in the middle. Indeed, this harks to a very similar debate in population genetics, between genetic drift and evolution by natural selection. Hubbell's neutral theory of communities is analogous to the processes of genetic drift at the population level. Our niche perspective is more aligned with the selectionist paradigm. Indeed, when it has been explored, very similar mechanisms are responsible for the maintenance of genetic and species diversity (Hastings and Gavrilets 1999; Amarasekare 2000), and thus the congruence between the fields should not be surprising. In the evolutionary literature, the relative importance of drift versus selection is certainly still an important issue, but synthetic models such as the "nearly neutral" model of population genetics (e.g., Ohta 1992) recognize the importance of both processes. Already, several models have begun to incorporate both niche and drift processes to make a number of important predictions (e.g., Loreau and Mouquet 1999; Amarasekare and Nisbet 2001; Chave et al. 2002; Mouquet and Loreau 2002; Wang et al. 2002), and experimental approaches considering both dispersal and niche differences have yielded interesting results (Tilman 1997; Shurin 2000, 2001; Foster 2001; Forbes and Chase 2002). We expect that the general conclusion from these and future theoretical models and empirical approaches will be that even when the processes of drift are important, the processes of trade-offs and niche differentiation are not eliminated. They simply must be examined within a slightly more complex framework.

11.4. Prospects

Ecology is in the midst of rapid change as a field of science. New theoretical and statistical tools are constantly being developed, such as computer-intensive individual-based models (IBMs) and quantitative methods of time series analysis in population dynamics. Additionally there is a steady stream of new empirical techniques (e.g., stable isotopes, GIS, atmospheric gases, paleoecological data linked to climate change). Often these methods appear to be almost independent of one another when in reality they are conceptually similar. Perhaps the reason these tools have not been better synthesized is the extremely complicated nature of each. As a result, ecologists are increasingly becoming specialists, not only within the subdisciplines of evolutionary, population, community, and ecosystems ecology, but also on the tools they use and the types of systems they study. A synthetic framework is critically needed to hold all this together. However, this framework needs to be rigorous enough that it does not lead to fundamental problems and flexible enough that it can illuminate connections among these approaches. We believe that the niche concept we have described in this book is such a framework and when combined, as a complement or comparison, with other frameworks will continue to serve as the foundation for ecological studies.

While we have attempted to draw links to a number of different fields and areas of enquiry, there are many questions that we have only touched upon or not discussed at all. In addition, solutions to problems that ecologists, and society as a whole, are facing—such as global climate change, invasive species, restoration ecology, and conservation of rare species—are in sore need of a strong conceptual foundation and are directly related to the niche concept. We hope that readers will be inspired by our attempt at a broad synthesis. However, we also hope that readers will focus more on the ideas than the details. We hope to have created more questions than we have answered.

LITERATURE CITED

Abrams, P. A. 1975. Limiting similarity and the form of the competition coefficient. Theor. Pop. Biol. 8:356–375.

Abrams, P. A. 1977. Density-independent mortality and interspecific competition: a test of Pianka's niche overlap hypothesis. Am. Nat. 111:539–552.

Abrams, P. A. 1980. Consumer functional response and competition in consumer-resource systems. Theor. Pop. Biol. 17:80–102.

Abrams, P. A. 1983. The theory of limiting similarity. Annu. Rev. Ecol. Syst. 14:359–376.

Abrams, P. A. 1986. Character displacement and niche shift analyzed using consumer-resource models of competition. Theor. Pop. Biol. 29:107–160.

Abrams, P. A. 1988. Resource productivity–consumer species diversity: simple models of competition in spatially heterogeneous environments. Ecology 69:1418–1433.

Abrams, P. A. 1993. Effects of increased productivity on the abundances of trophic levels. Am. Nat. 141:351–371.

Abrams, P. A. 1995. Monotonic or unimodal diversity-productivity gradients: what does competition theory predict? Ecology 76:2019–2027.

Abrams, P. A. 1998. High competition with low similarity and low competition with high similarity: The interaction of exploitative and apparent competition in consumer-resource systems. Am. Nat. 152:114–128.

Abrams, P. A. 1999. Is predator mediated coexistence possible in unstable systems? Ecology 80:608–621.

Abrams, P. A. 2001a. Describing and quantifying interspecific interactions: a commentary on recent approaches. Oikos 94:209–218.

Abrams, P. A. 2001b. Modelling the adaptive dynamics of traits involved in inter- and intraspecific interactions: an assessment of three methods. Ecol. Letters 4:166–175.

Abrams, P. A., and L. R. Ginzburg. 2000. The nature of predation: prey dependent, ratio dependent, or neither? Trends Ecol. Evol. 15:337–341.

Abrams, P. A., and R. D. Holt. 2002. The impact of consumer-resource cycles on the coexistence of competing consumers. Theor. Pop. Biol. 62:281–295.

Abrams, P. A., and J. Roth. 1994. The responses of unstable food chains to enrichment. Evol. Ecol. 8:150–171.

Abramsky, Z., M. L. Rosenzweig, and A. Subach. 1997. Safety in numbers: sophisticated vigilance by Allenby's gerbil. Proc. Nat. Acad. Sci. (USA) 94:5713–5715.

Abramsky, Z., M. L. Rosenzweig, and A. Subach. 1998. Do gerbils care more about competition or predation? Oikos 83:75–84.

Abramsky, Z., M. L. Rosenzweig, and A. Subach. 2000. The energetic cost of competition: gerbils as moneychangers. Evol. Ecol. Res. 2:279–292.

Abramsky, Z., M. L. Rosenzweig, and B. Pinshow. 1991. The shape of a gerbil isocline: an experimental field study using principles of optimal habitat selection. Ecology 72:329–340.

Abramsky, Z., O. Ovadia, and M. L. Rosenzweig. 1994. The shape of a *Gerbillus pyramidum* (Rodentia, Gerbillinae) isocline: an experimental field study. Oikos 69:318–326.

Agrawal, A. A. 2001. Phenotypic plasticity in the interactions and evolution of species. Science 294:321–326.

Agrawal, A. A., S. Y. Strauss, and M. J. Stout. 1999. Costs of induced responses and tolerance to herbivory in male and female fitness components of wild radish. Evolution 53:1093–1104.

Alroy, J., C. R. Marshall, R. K. Bambach, K. Bezuko, M. Foote, F. T. Fulsich, T. A. Hansen, S. M. Holland, L. C. Ivory, D. Jablonski, D. K. Jacobs, D. C. Jones, M. A. Kosnik, S. Lidgard, S. Low, A. I. Miller, P. M. Novack-Gottshall, T. D. Olszewski, M. E. Patzkowsky, D. M. Raup, K. Roy, J. J. Seposki, M. G. Sommers, P. J. Wagner, and A. Webber. 2001. Effects of sampling standardization on estimates of Phanerozoic marine diversification. Proc. Nat. Acad. Sci. (USA) 98:6261–6266.

Amarasekare, P. 2000. The geometry of coexistence. Biol. J. Linnean Soc. 71:1–31.

Amarasekare, P., and R. M. Nisbet. 2001. Spatial heterogeneity, source-sink dynamics, and the local coexistence of competing species. Am. Nat. 158:572–584.

Arditi, R., and L. R. Ginzburg. 1989. Coupling in predator-prey dynamics: ratio-dependence. J. Theor. Biol. 139:311–326.

Armstrong, R. A., and R. McGehee. 1976. Coexistence of species competing for shared resources. Theor. Pop. Biol. 9:317–328.

Armstrong, R. A., and R. McGehee. 1980. Competitive exclusion. Am. Nat. 115:151–170.

Balmford, A., A. Bruner, P. Cooper, R. Costanza, S. Farber, R. E. Green, M. Jenkins, P. Jefferiss, V. Jessamy, J. Madden, K. Munro, N. Myers, S. Naeem, J. Paavola, M. Rayment, S. Rosendo, J. Roughgarden, K. Trumper, and R. K. Turner. 2002. Economic reasons for conserving wild nature. Science 297:950–953.

Balvanera, P., G. C. Daily, P. R. Ehrlich, T. H. Ricketts, S. Kark, C. Kremen, and H. Pereira. 2001. Conserving biodiversity and ecosystem services. Science 291:2047.

Bazely, D. R., and R. L. Jeffries. 1986. Changes in the composition and standing crop of salt-marsh communities in response to the removal of a grazer. J. Ecol. 74:693–706.

Begon, M., J. Harper, and C. Townsend. 1996. Ecology: individuals, populations, and communities. 3d ed. Oxford: Blackwell.

Bell, G. 1982. The masterpiece of nature: the evolution and genetics of sexuality. Berkeley and Los Angeles: University of California Press.

Bell, G. 2000. The distribution of abundance in neutral communities. Am. Nat. 155: 606–617.

Bell, G. 2001. Neutral macroecology. Science 293:2413–2418.

Bender, E. A., T. J. Case, and M. E. Gilpin. 1984. Perturbation experiments in community ecology: theory and practice. Ecology 65:1–13.

Bergelson, J., and C. B. Purrington. 1996. Surveying patterns in the costs of resistance in plants. Am. Nat. 148:536–558.

Berlow, E. L., S. A. Navarette, C. J. Briggs, and M. E. Power. 1999. Quantifying variation in the strengths of species interactions. Ecology 80:2206–2224.

Bohannan, B. J. M., and R. E. Lenski. 1997. Effect of resource enrichment on a chemostat community of bacteria and bacteriophage. Ecology 78:2303–2315.

Bohannan, B. J. M., and R. E. Lenski. 1999. Effect of prey heterogeneity on the response of a model food chain to resource enrichment. Am. Nat. 153:73–82.

Bohannan, B. J. M., and R. E. Lenski. 2000. The relative importance of competition and predation varies with productivity in a model community. Am. Nat. 156:329–340.

Bond, E., and J. M. Chase. 2002. Local and regional controls of ecosystem function. Ecol. Letters 5:467–470.

Bovbjerg, R. V. 1970. Ecological isolation and competitive exclusion in two crayfish (*Orconectes virilis* and *Orconectes immunis*). Ecology 51:225–236.

Bowers, M. A., and J. H. Brown. 1982. Body size and coexistence in desert rodents: chance or community structure? Ecology 63:391–400.

Brokaw, N., and R. T. Busing. 2000. Niche versus chance and tree diversity in forest gaps. Trends Ecol. Evol. 15:183–188.

Brown, J. H. 1981. Two decades of homage to Santa Rosalia: towards a general theory of biodiversity. Am. Zool. 21:877–888.

Brown, J. H. 1995. Macroecology. Chicago: University of Chicago Press.

Brown, J. H. 1997. An ecological perspective on the challenge of complexity. EcoEssay Series no. 1. National Center for Ecological Analysis and Synthesis, www.nceas.ucsb.edu.

Brown, J. H. 1998. The granivory experiments at Portal. In Experimental ecology: issues and perspectives, ed. W. J. Resetarits and J. Bernardo, 71–95. Oxford: Oxford University Press.

Brown, J. H., B. J. Fox, and D. A. Kelt. 2000. Assembly rules: desert rodent communities are structured at scales from local to regional. Am. Nat. 156:314–321.

Brown, J. S. 1988. Patch use as an indicator of habitat preference, predation risk, and competition. Behav. Ecol. Sociobiol. 22:37–47.

Brown, J. S. 1989. Desert rodent community structure: a test of four mechanisms of coexistence. Ecol. Monogr. 59:1–20.

Brown, J. S. 1992. Patch use under predation risk. Part 1. Models and predictions. Ann. Zool. Fennici 29:301–309.

Brown, J. S. 1996. Coevolution and community organization in three habitats. Oikos 75:193–206.

Brown, J. S. 1998. Game theory and habitat selection. In Game theory and animal behavior, ed. L. A. Dugatkin and H. K. Reeve, 188–220. Oxford: Oxford University Press.

Brown, J. S. 1999. Vigilance, patch use, and habitat selection: foraging under predation risk. Evol. Ecol. Res. 1:49–71.

Brown, J. S., and T. L. Vincent. 1987. Coevolution as an evolutionary game. Evolution 41:66–79.

Brown, J. S., and T. L. Vincent. 1992. Organization of predator-prey communities as an evolutionary game. Evolution 46:1269–1283.

Brown, J. S., B. P. Kotler, and W. A. Mitchell. 1994. Foraging theory, patch use, and the structure of a Negev desert granivore community. Ecology 75:2286–2300.

Brown, W. L., Jr., and E. O. Wilson. 1956. Character displacement. Syst. Zool. 5:49–64.

Buckland, S. M., and J. P. Grime. 2000. The effects of trophic structure and soil fertility on the assembly of plant communities: a microcosm experiment. Oikos 91:336–352.

Buckling, A., R. Kassen, G. Bell, and P. B. Rainey. 2000. Disturbance and diversity in experimental microcosms. Nature 408:961–964.

Caceres, C. E. 1997. Temporal variation, dormancy, and coexistence: a field test of the storage effect. Proc. Nat. Acad. Sci. (USA) 94:9171–9175.

Caraco, T. 1980. On foraging time allocation in a stochastic environment. Ecology 61: 119–128.

Carpenter, S. R. 1996. Microcosm experiments have limited relevance for community and ecosystem ecology. Ecology 77:677–680.

Carpenter, S. R., and J. F. Kitchell. 1993. The trophic cascade in lakes. Cambridge: Cambridge University Press.

Carpenter, S. R., J. J. Cole, J. R. Hodgson, J. F. Kitchell, M. L. Pace, D. Bade, K. L. Cottingham, T. E. Essington, J. N. Houser, and D. E. Schindler. 2001. Trophic cascades, nutrients, and lake productivity: whole-lake experiments. Ecol. Monogr. 71: 163–186.

Case, T. J., and M. L. Taper. 2000. Interspecific competition, gene flow, and the coevolution of species boarders. Am. Nat. 155:583–605.

Caswell, H. 1976. Community structure: a neutral model analysis. Ecol. Monogr. 46: 327–354.

Caswell, H. 2001. Matrix population models: construction, analysis, and interpretation. Sunderland, Mass.: Sinauer Associates.

Chapin, F. S., E. S. Zavalera, V. T. Eviner, R. L. Naylor, P. M. Vitousek, H. L. Reynolds, D. U. Hooper, S. Lavorel, O. E. Sala, S. E. Hobbie, M. C. Mack, and S. Diaz. 2000. Consequences of changing biodiversity. Nature 405:234–242.

Charlesworth, B. 1990. Optimization models, quantitative genetics, and mutation. Evolution 44:520–538.

Charnov, E. L. 1976. Optimal foraging: the marginal value theorem. Theor. Pop. Biol. 9:129–136.

Chase, J. M. 1996a. Differential competitive interactions and the included niche: an experimental analysis with grasshoppers. Oikos 76:103–112.

Chase, J. M. 1996b. Varying resources and competitive dynamics. Am. Nat. 147:649–654.

Chase, J. M. 1999a. Food web effects of prey size-refugia: variable interactions and alternative stable equilibria. Am. Nat. 154:559–570.

Chase, J. M. 1999b. To grow or reproduce? the role of life-history plasticity in food webs. Am. Nat. 154:571–586.

Chase, J. M. 2000. Are there real differences among aquatic and terrestrial food webs? Trends Ecol. Evol. 15:408–412.

Chase, J. M., and T. M. Knight. 2003. Community genetics: towards a synthesis. Ecology 84: (in press).

Chase, J. M., and M. A. Leibold. 2002. Spatial scale dictates the productivity-diversity relationship. Nature 416:427–430.

Chase, J. M., M. A. Leibold, A. L. Downing, and J. B. Shurin. 2000. The effects of productivity, herbivory and plant species turnover in grassland food webs. Ecology 81:2485–2497.

Chase, J. M., M. A. Leibold, and E. L. Simms. 2000. Plant tolerance and resistance in food webs: community-level predictions and evolutionary implications. Evol. Ecol. 14:289–314.

Chase, J. M., W. G. Wilson, and S. A. Richards. 2001. Foraging trade-offs and resource patchiness: theory and experiments with a freshwater snail community. Ecol. Letters 4:304–312.

Chave, J., H. Muller-Landau, and S. A. Levin. 2002. Comparing classical community models: theoretical consequences for patterns of diversity. Am. Nat. 159:1–23.

Chesson, P. 1985. Coexistence of competitors in spatially and temporally varying environments: a look at the combined effects of different sorts of variability. Theor. Pop. Biol. 28:263–287.

Chesson, P. 1990. MacArthur's consumer-resource model. Theor. Pop. Biol. 37:26–38.

Chesson, P. 1991. A need for niches? Trends Ecol. Evol. 6:26–28.

Chesson, P. 1994. Multispecies competition in variable environments. Theor. Pop. Biol. 45:227–276.

Chesson, P. 2000a. General theory of competitive coexistence in spatially-varying environments. Theor. Pop. Biol. 58:211–237.

Chesson, P. 2000b. Mechanisms of maintenance of species. Annu. Rev. Ecol. Syst. 31: 343–366.

Chesson, P., and N. Huntly. 1997. The role of harsh and fluctuating conditions in the dynamics of ecological communities. Am. Nat. 150:519–553.

Chesson, P., and R. R. Warner. 1981. Environmental variability promotes coexistence in lottery competitive systems. Am. Nat. 117:923–943.

Chesson, P., S. Pacala, and C. Neuhauser. 2002. Environmental niches and ecosystem functioning. In The functional consequences of biodiversity: empirical progress and theoretical extensions, ed. A. E. Kinzig, S. W. Pacala, and D. Tilman, 213–245. Princeton: Princeton University Press.

Clements, F. E. 1919. Plant succession. Carnegie Institution of Washington publication 212. Washington: Carnegie Institution of Washington.

Clements, F. E. 1936. Nature and structure of the climax. J. Ecol. 24:252–282.

Cochran-Stafira, D. L., and C. N. von Ende. 1998. Integrating bacteria into food webs: studies with *Sarracenia purpurea* inquilines. Ecology 79:880–898.

Cody, M. L. 1973. Character convergence. Annu. Rev. Ecol. Syst. 4:189–211.

Colwell, R. K. 1992. Niche: a bifurcation in the conceptual lineage of the term. In Keywords in evolutionary biology, ed. E. Fox-Keller and E. A. Loyd, 241–248. Cambridge: Harvard University Press.

Colwell, R. K., and D. J. Futuyma. 1971. On the measurement of niche breadth and overlap. Ecology 52:567–576.

Colwell, R. K., and D. W. Winkler. 1984. A null model for null models in biogeography. In Ecological communities: conceptual issues and the evidence, ed. D. R. Strong, D. S. Simberloff, L. G. Abele, and A. B. Thistle, 344–359. Princeton: Princeton University Press.

Colwell, R. K., and E. R. Fuentes. 1975. Experimental studies of the niche. Annu. Rev. Ecol. Syst. 6:281–310.

Condit, R., N. Pitman, E. G. Leigh Jr., J. Chave, J. Terborgh, R. B. Foster, P. Núñez, S. Aguilar, R. Valencia, G. Villa, H. C. Muller-Landau, E. Losos, and S. P. Hubbell. 2002. Beta-diversity in tropical forest trees. Science 295:666–669.

Connell, J. H. 1961a. Effects of competition, predation by *Thais lapillus*, and other factors on natural populations of the barnacle, *Balanus balanoides*. Ecol. Monogr. 31:61–104.

Connell, J. H. 1961b. The influence of interspecific competition and other factors on the distribution of the barnacle *Chthamalus stellatus*. Ecology 42:710–723.

Connell, J. H. 1975. Some mechanisms producing structure in natural communities: a model and evidence from field experiments. In Ecology and evolution of communities, ed. M. L. Cody and J. M. Diamond, 460–490. Cambridge: Harvard University Press.

Connell, J. H. 1978. Diversity in tropical rainforests and coral reefs. Science 199:1302–1310.

Connell, J. H. 1983. On the prevalence and relative importance of interspecific competition: evidence from field experiments. Am. Nat. 122:661–696.

Connell, J. H., and R. O. Slatyer. 1977. Mechanisms of succession in natural communities and their role in community stability and organization. Am. Nat. 111:1119–1144.

Connor, E. F., and D. S. Simberloff. 1978. Species number and compositional similarity of the Galapagos flora and avifauna. Ecol. Monogr. 48:219–248.

Connor, E. F., and D. S. Simberloff. 1979. The assembly of species communities. Chance or competition? Ecology 60:1132–1140.

Connor, E. F., and D. S. Simberloff. 1984. Neutral models of species co-occurrence patterns and Rejoinders. In Ecological communities: conceptual issues and the evidence, ed. D. R. Strong, D. S. Simberloff, L. G. Abele, and A. B. Thistle, 316–331, 341–343. Princeton: Princeton University Press.

Costanza, R., R. d'Arge, R. de Groot, S. Farber, M. Grasso, B. Hannon, K. Limburg, S. Naeem, R. V. O'Neill, J. Paruelo, R. G. Raskin, P. Sutton, M. VandenBelt. 1996. The value of the world's ecosystem services and natural capital. Nature 387:253–260.

Cowles, H. C. 1899. The ecological relations of the vegetation on the sand dunes of Lake Michigan. Bot. Gaz. 27:95–117, 167–202, 281–308, 361–391.

Cramer, N. F., and R. M., May. 1971. Interspecific competition, predation, and species diversity: a comment. J. Theor. Biol. 34:289–293.

Currie, D. J. 1991. Energy and large-scale patterns of animal and plant species richness. Am. Nat. 137:27–49.

Daily, G. C., T. Soderqvist, S. Aniyar, K. Arrow, P. Dasgupta, P. R. Ehrlich, C. Folke, A. Jansson, B. O. Jansson, N. Kautsky, S. Levin, J. Lubchenco, K. G. Maler, D. Simpson, D. Starett, D. Tilman, and B. Walker. 2000. Ecology: the value of nature and the nature of value. Science 289:395–396.

Darwin, C. R. 1859. The origin of species by means of natural selection, or the preservation of favoured races in the struggle for life. London: John Murray.

Davis, M. B. 1981. Quaternary history and the stability of forest communities. In Forest succession: concepts and application, ed. D. C. West, H. H. Shugart, and D. B. Botkin, 132–153. New York: Springer-Verlag, New York.

Davis, M. B. 1986. Climatic instability, time lags, and community disequilibrium. In Community ecology, ed. J. M. Diamond and T. J. Case, 269–284. New York: Harper and Row.

Day, T., P. A. Abrams, and J. M. Chase. 2002. Some effects of size-specific predation on life-history evolution. Evolution 56:877–887.

Dayan, T., and D. S. Simberloff. 1994. Character displacement, sexual dimorphism, and morphological variation among British and Irish mustelids. Ecology 75:1063–1073.

Dayan, T., D. S. Simberloff, E. Tchernov, and Y. Yom-Tov. 1990. Feline canines: Community-wide character displacement among the small cats of Israel. Am. Nat. 136:39–60.

Dayton, P. K. 1971. Competition, disturbance, and community organization: the provision and subsequent utilization of space in a rocky intertidal community. Ecol. Monogr. 41:351–389.

de Mazancourt, C., and M. Loreau. 2000. Effect of herbivory and plant species replacement on primary production. Am. Nat. 155:735–754.

DeAngelis, D. 1992. Dynamics of nutrient cycling and food webs. London: Chapman and Hall.

Diamond, J. M. 1975. Assembly of species communities. In Ecology and evolution of communities, ed. M. L. Cody and J. M. Diamond, 342–444. Cambridge: Harvard University Press.

Dieckmann, U., and M. Doebeli. 1999. On the origin of species by sympatric speciation. Nature 400:354–357.

Dieckmann, U., R. Law, and J. A. J. Metz, eds. 2000. The geometry of ecological interactions: simplifying spatial complexity. Cambridge: Cambridge University Press.

Diehl, S., and M. Feißel. 2000. Effects of enrichment on three-level food chains with omnivory. Am. Nat. 155:200–218.

Doak, D. F., D. Bigger, E. K. Harding, M. A. Marvier, R. E. O'Malley, and D. Thomson.

1998. The statistical inevitability of stability-diversity relationships in community ecology. Am. Nat. 151:264–276.

Dobzhansky, T. 1937. Genetics and the origin of species. New York: Columbia University Press.

Dodson, S. I. 1970. Complementary feeding niches sustained by size-selective predation. Limnol. Oceanogr. 15:131–137.

Dodson, S. I., S. E. Arnott, and K. L. Cottingham. 2000. The relationship in lake communities between primary productivity and species richness. Ecology 81:2662–2679.

Downing, A. L., and M. A. Leibold. 2002. Ecosystem consequences of species richness and composition in pond food webs. Nature 416:837–841.

Drake, J. A. 1990. The mechanics of community assembly and succession. J. Theor. Biol. 147:213–233.

Drake, J. A. 1991. Community assembly mechanics and the structure of an experimental species ensemble. Am. Nat. 137:1–26.

Drake, J. A., C. R. Zimmermann, T. Purucker, and C. Rojo. 1999. On the nature of the assembly trajectory. In Ecological assembly rules: perspectives, advances, retreats, ed. E. Weiher and P. A. Keddy, 233–250. Cambridge: Cambridge University Press.

Drake, J. A., G. R. Huxel, and C. L. Hewitt. 1996. Microcosms as models for generating and testing community theory. Ecology 77:670–677.

Drake, J. A., T. E. Flum, G. J. Witteman, T. Voskuil, A. M. Hoylman, C. Creason, D. A. Kenny, G. R. Huxel, C. S. Larue, and J. R. Duncan. 1993. The construction and assembly of an ecological landscape. J. Anim. Ecol. 62:117–130.

Ebenman, B., and L. Persson. 1988. Size-structured populations: ecology and evolution. New York: Springer-Verlag.

Egler, F. E. 1954. Vegetation science concepts. Part 1. Initial floristic composition–factor in old-field vegetation development. Vegetatio 4:412–417.

Elton, C. 1927. Animal ecology. London: Sidgwick and Jackson.

Enquist, B. J., J. P. Haskell, and B. H. Tiffney. 2002. General patterns of taxonomic and biomass partitioning in extant and fossil plant communities. Nature 419:610–613.

Ernest, S. K. M., and J. H. Brown. 2001. Delayed compensation for missing keystone species by colonization. Science 292:101–104.

Fisher, R. A. 1942. The genetical theory of natural selection. New York: Dover.

Forbes, A. E., and J. M. Chase. 2002. The role of habitat connectivity and landscape geometry on experimental zooplankton metacommunities. Oikos 96:433–440.

Foster, B. L. 2001. Constraints on colonization and species richness along a grassland productivity gradient: the role of propagule availability. Ecol. Letters 4:530–535.

Fox, J. W. 2002. Testing a simple rule for dominance in resource competition. Am. Nat. 159:305–319.

Frost, T. M., P. K. Montz, T. K. Kratz, T. Badillo, P. L. Brezonik, M. J. Gonzales, C. J. Watras, K. E. Webster, J. G. Weiner, C. E. Williamson, and D. P. Morris. 1999. Multiple stresses from a single agent: diverse responses to the experimental acidification of Little Rock Lake, Wisconsin. Limnol. Oceanogr. 44:784–794.

Fryxell, J. M., and P. Lundberg. 1997. Individual behavior and community dynamics. New York: Chapman and Hall.

Garbutt, K., and A. R. Zangerl. 1983. Application of genotype-environment interaction analysis to niche quantification. Ecology 64:1292–1296.

Gaston, K. J. 2000. Global patterns in biodiversity. Nature 405:220–227.

Gaston, K. J., and T. M. Blackburn. 2000. Pattern and process in macroecology. Oxford: Blackwell.

Gause, G. F. 1936. The struggle for existence. Baltimore: Williams and Wilkins.
Geisel, T. S. [Dr. Seuss]. 1955. On beyond zebra. New York: Random House.
Gersani, M., J. S. Brown, E. E. O'Brien, G. M. Maina, and Z. Abramsky. 2001. Tragedy of the commons as a result of root competition. J. Ecol. 89:660–669.
Giller, P. J. 1984. Community structure and the niche. London: Chapman and Hall.
Gilliam, J. F., and D. F. Fraser. 1987. Habitat selection under predation hazard: test of a model with foraging minnows. Ecology 68:1856–1862.
Gilpin, M. E., and J. M. Diamond. 1984. Are species co-occurrences on islands non-random, and are null hypotheses useful in community ecology; with rejoinders. In Ecological communities: conceptual issues and the evidence, ed. D. R. Strong, D. S. Simberloff, L. G. Abele, and A. B. Thistle, 297–315, 332–341. Princeton: Princeton University Press.
Gilpin, M. E., and T. J. Case. 1976. Multiple domains of attraction in competition communities. Nature 261:40–42.
Gleason, H. A. 1917. The structure and development of plant association. Bull. Torrey Bot. Club 44:463–481.
Gleason, H. A. 1926. The individualistic concept of plant association. Bull. Torrey Bot. Club 53:7–26.
Gleason, H. A. 1927. Further views on the succession concept. Ecology 8:299–326.
Gleeson, S. K. 1994. Density-dependence is better than ratio-dependence. Ecology 75: 1834–1835.
Gleeson, S. K., and D. Tilman. 1992. Plant allocation and the multiple limitation hypothesis. Am. Nat. 139:1322–1343.
Godfray, H. C. J., and J. H. Lawton. 2001. Scale and species numbers. Trends Ecol. Evol. 16:400–404.
Goldberg, D. E., and T. E. Miller. 1990. Effects of different resources additions on species diversity in an annual plant community. Ecology 71:213–225.
Goldwasser, L., J. Cook, and E. D. Silverman. 1994. The effects of variability on metapopulation dynamics and rates of invasion. Ecology 75:40–47.
Gomulkiewicz, R., J. N. Thompson, R. D. Holt, S. L. Nuismer, and M. E. Hochberg. 2000. Hot spots, cold spots, and the geographic mosaic theory of coevolution. Am. Nat. 156:156–174.
Gomulkiewicz, R., R. D. Holt, and M. Barfield. 1999. The effects of density-dependence and immigration on local adaptation in a black-hole sink environment. Theor. Pop. Biol. 55:283–296.
Gotelli, N. J., and G. R. Graves. 1996. Null models in ecology. Washington: Smithsonian Institution Press.
Gotelli, N. J., and D. J. McCabe. 2002. Species co-occurrence: a meta-analysis of J. M. Diamond's assembly rules model. Ecology 83:2091–2096.
Grace, J. B., and R. G. Wetzel. 1981. Habitat partitioning and competitive displacement in cattails (Typha): experimental field studies. Am. Nat. 118:463–474.
Graham, M. H., and P. K. Dayton. 2002. On the evolution of ecological ideas: paradigms and scientific progress. Ecology 83:1481–1489.
Grant, P. R. 1986. Ecology and evolution of Darwin's finches. Princeton: Princeton University Press.
Griesemer, J. R. 1992. Niche: Historical perspectives. In Keywords in evolutionary biology, ed. E. Fox-Keller and E. A. Loyd, 231–240. Cambridge: Harvard University Press.
Grinnell, J. 1904. The origin and distribution of the chestnut-backed chickadee. Auk 21:375–377.
Grinnell, J. 1914. An account of the mammals and birds of the lower Colorado River Valley. University of California Publications in Zoology 12:51–294.

Grinnell, J. 1917. The niche-relationships of the California thrasher. Auk 34:427–433.
Grinnell, J. 1924. Geography and evolution. Ecology 5:225–229.
Grinnell, J., and H. S. Swarth. 1913. An account of the birds and mammals of the San Jacinto area of southern California. University of California Publications in Zoology 10:197–406.
Gross, K. L., M. R. Willig, L. Gough, R. Inouye, and S. B. Cox. 2000. Patterns of species density and productivity at different spatial scales in herbaceous plant communities. Oikos 89:417–427.
Grover, J. P. 1988. Dynamics of competition within a variable environment: experiments with two diatom species. Ecology 69:408–417.
Grover, J. P. 1990. Resource competition in a variable environment: phytoplankton growing according to Monod's model. Am. Nat. 136:771–789.
Grover, J. P. 1991. Resource competition among microalgae in variable environments: experimental tests of alternative models. Oikos 62:231–243.
Grover, J. P. 1994. Assembly rules for communities for nutrient limited plants and specialist herbivores. Am. Nat. 143:258–282.
Grover, J. P. 1995. Competition, herbivory, and enrichment: nutrient-based models for edible and inedible plants. Am. Nat. 145:746–774.
Grover, J. P. 1997. Resource competition. London: Chapman Hall.
Grover, J. P., and Holt, R. D. 1998. Disentangling resource and apparent competition: realistic models for plant-herbivore communities. J. Theor. Biol. 191:353–376.
Grover, J. P., and J. H. Lawton. 1994. Experimental studies on community convergence and alternative stable states: comments on a paper by Drake et al. J. Anim. Ecol. 63:484–487.
Grubb, P. 1977. The maintenance of species richness in plant communities: the importance of the regeneration niche. Biol. Rev. 52:107–145.
Gurevitch, J., J. A. Morrison, and L. V. Hedges. 2000. The interaction between competition and predation: a meta-analysis of field experiments. Am. Nat. 155:435–453.
Gurevitch, J., L. L. Morrow, A. Wallace, and J. J. Walsh. 1992. A meta-analysis of competition in field experiments. Am. Nat. 140:539–572.
Gurney, W. S. C. and R. M. Nisbet. 1998. Ecological dynamics. Oxford University Press.
Hairston, N. G., Jr. 1995. Commentary. Ecology 76:1371.
Hall, D. J., W. E. Cooper, and E. E. Werner. 1970. An experimental approach to the production dynamics and structure of freshwater animal communities. Limnol. Oceanogr. 15:839–928.
Hansen, S. R., and S. P. Hubbell. 1980. Single nutrient microbial competition: qualitative agreement between experimental and theoretically forecasted outcomes. Science 207:1491–1493.
Hardin, G. 1960. The competitive exclusion principle. Science 131:1292–1297.
Harvey, P. H., R. K., Colwell, J. W., Silvertown, and R. M. May. 1983. Null models in ecology. Annu. Rev. Ecol. Syst. 14:189–211.
Hassell, M. P. 1978. The dynamics of arthropod predator-prey systems. Princeton: Princeton University Press.
Hastings, A. 1980. Disturbance, coexistence, history, and competition for space. Theor. Pop. Biol. 18:363–373.
Hastings, A., and S. Gavrilets. 1999. Global dispersal reduces local diversity. Proc. Roy. Soc. London (B) 266:2067–2070.
Holt, R. D. 1977. Apparent competition and the structure of prey communities. Theor. Pop. Biol. 12:197–229.
Holt, R. D. 1984. Spatial heterogeneity, indirect interactions, and the coexistence of prey species. Am. Nat. 124:377–406.

Holt, R. D. 1985. Density-independent mortality, non-linear competitive interactions, and species coexistence. J. Theor. Biol. 116:479–493.

Holt, R. D. 1987. Prey communities in patchy environments. Oikos 50:276–290.

Holt, R. D. 1996. Adaptive evolution in source-sink environments: direct and indirect effects of density-dependence on niche evolution. Oikos 75:182–192.

Holt, R. D. 1997. Community modules. In Multitrophic interactions in terrestrial systems, ed. A. C. Grange and V. K. Brown, 333–350. Oxford: Blackwell.

Holt, R. D., and M. S. Gaines. 1992. The analysis of adaptation in heterogeneous landscapes: implications for the evolution of fundamental niches. Evol. Ecol. 6:433–447.

Holt, R. D., and R. Gomulkiewicz. 1997. How does immigration influence local adaptation? A reexamination of a familiar paradigm. Am. Nat. 149:563–572.

Holt, R. D., J. Grover, and D. Tilman. 1994. Simple rules for interspecific dominance in systems with exploitation and apparent competition. Am. Nat. 144:741–771.

Holt, R. D., and J. H. Lawton. 1993. Apparent competition and enemy-free space in insect host-parasitoid communities. Am. Nat. 142:623–645.

Holt, R. D., and M. Loreau. 2002. Biodiversity and ecosystem functioning: the role of trophic interactions and the importance of system openness. In The functional consequences of biodiversity: empirical progress and theoretical extensions, ed. A. E. Kinzig, S. W. Pacala, and D. Tilman, 246–262. Princeton: Princeton University Press.

Holt, R. D., and G. A. Polis. 1997. A theoretical framework for intraguild predation. Am. Nat. 149:745–764.

Hooper, D. U., and P. M. Vitousek. 1997. The effects of plant composition and diversity on ecosystem processes. Science 277:1302–1305.

Horn, H. S., and R. H. MacArthur. 1972. Competition among fugitive species in a harlequin environment. Ecology 53:749–752.

Hubbell, S. P. 1979. Tree dispersion, abundance, and diversity in a tropical dry forest. Sciences 203:1299–1303.

Hubbell, S. P. 1997. A unified theory of biogeography and relative species abundance its application to tropical rain forests and coral reefs. Coral Reefs 16, suppl.:S9–S21.

Hubbell, S. P. 2001. The unified neutral theory of species abundance and diversity. Princeton: Princeton University Press.

Hubbell, S. P., and R. B. Foster. 1986. Biology, chance, and history and the structure of tropical rainforest communities. In Community ecology, ed. J. M. Diamond and T. J. Case, 314–329. New York: Harper and Row.

Hubbell, S. P., R. B. Foster, S. T. O'Brien, K. E. Harms, R. Condit, B. Wechsler, S. J. Wright, and S. Loo de Lao. 1999. Light-gap disturbances, recruitment limitation, and tree diversity in a Neotropical forest. Science 283:554–557.

Huisman, J., and F. J. Weissing. 1995. Competition for nutrients and light in a mixed water column: a theoretical analysis. Am. Nat. 146:536–564.

Huisman, J., and F. J. Weissing. 1999. Biodiversity of plankton by species oscillations and chaos. Nature 402:407–410.

Huisman, J., and F. J. Weissing. 2001a. Biological conditions for oscillations and chaos generated by multispecies competition. Ecology 82:2682–2695.

Huisman, J., and F. J. Weissing. 2001b. Fundamental unpredictability in multispecies competition. Am. Nat. 157:488–494.

Huisman, J., J. P. Grover, R. van der Wal, and J. van Andel. 1999. Competition for light, plant species replacement, and herbivore abundance along productivity gradients. In Herbivores: between plants and predators, ed. H. Olff, V. K. Brown, and R. H. Drent, 239–270. Oxford: Blackwell.

Hunter, M. D., and P. W. Price. 1992. Playing chutes and ladders: heterogeneity and the relative roles of bottom-up and top-down forces in natural communities. Ecology 73:724–732.

Hurlbert, S. H. 1981. A gentle depilation of the niche: Dicean resource sets in hyperspace. Evol. Theory 5:177–184.

Huston, M. A. 1979. A general hypothesis of species diversity. Am. Nat. 113:81–101.

Huston, M. A. 1994. Biological diversity: the coexistence of species in changing landscapes. Cambridge: Cambridge University Press.

Huston, M. A. 1997. Hidden treatments in ecological experiments: re-evaluating the ecosystem function of biodiversity. Oecologia 110:449–460.

Huston, M. A. 1999. Local processes and regional patterns: appropriate scales for understanding variation in the diversity of plants and animals. Oikos 86:393–401.

Huston, M. A., L. W. Arssen, M. P. Austin, B. S. Cade, J. D. Fridley, E. Garnier, J. P. Grime, J. Hodgson, W. K. Lauenroth, K. Thomson, J. H. Vandermeer, and D. A. Wardle. 2000. No consistent effect of plant diversity on productivity. Science 289: 1255.

Hutchinson, G. E. 1944. Limnological studies in Connecticut. Part 7. A critical examination of the supposed relationship between phytoplankton periodicity and chemical changes in lake waters. Ecology 25:3–26.

Hutchinson, G. E. 1957. Concluding remarks. Cold Springs Harbor Symp. Quant. Biol. 22:415–427.

Hutchinson, G. E. 1959. Homage to Santa Rosalia, or why are there so many kinds of animals? Am. Nat. 93:145–159.

Hutchinson, G. E. 1961. The paradox of the plankton. Am. Nat. 95:137–145.

Hutchinson, G. E. 1965. The ecological theater and the evolutionary play. New Haven: Yale University Press.

Hutchinson, G. E. 1978. An introduction to population ecology. New Haven: Yale University Press.

Huxley, J. S. 1942. Evolution: the modern synthesis. New York: Harper.

Inouye, R. S., and D. Tilman. 1995. Convergence and divergence of old-field vegetation after 11-years of nitrogen addition. Ecology 76:1872–1887.

Ives, A. R., and J. B. Hughes. 2002. General relationships between species diversity and stability in competitive systems. Am. Nat. 159:388–395.

Ives, A. R., J. L. Klug, and K. Gross. 2000. Stability and species richness in complex communities. Ecol. Letters 3:399–411.

Jablonski, D., and J. J. Sepkoski Jr. 1996. Paleobiology, community ecology, and scales of ecological pattern. Ecology 1367–1378.

Jablonski, D., J. J. Sepkoski Jr., D. J. Botther, and P. M. Sheehan. 1983. Onshore-offshore patterns in the evolution of Phanerozoic shelf communities. Science 222: 1123–1125.

Jackson, J. B. C. 1981. Interspecific competition and species distributions: the ghosts of theories and data past. Am. Zool. 21:889–901.

Jackson, J. B. C., A. F. Budd, and J. M. Pandolfi. 1996. The shifting balance of natural communities? In Evolutionary paleobiology, ed. D. Jablonski, D. Erwin, and J. H. Lipps, 89–127. Chicago: University of Chicago Press.

Jackson, J. B. C., and K. G. Johnson. 2000. Life in the last few million years. Paleobiology 26:221–235.

Janzen, D. H. 1970. Herbivores and the number of trees in tropical forests. Am. Nat. 104:501–528.

Jenkins, D. G., and A. L. Buikema Jr. 1998. Do similar communities develop in similar sites? A test with zooplankton structure and function. Ecol. Monogr. 68:421–443.

Johnson, R. H. 1910. Determinant evolution in the color pattern of the lady-beetles. Washington: Carnegie Institution of Washington.

Jones, C. G., and J. H. Lawton, eds. 1995. Linking species and ecosystems. London: Chapman and Hall.

Jones, C. G., J. H. Lawton, and M. Shachak. 1997. Positive and negative effects of organisms as physical ecosystem engineers. Ecology 78:1946–1957.

Jonsson, B. G. 2001. A null model for randomization tests of nestedness in species assemblages. Oecologia 127:309–313.

Juenger, T., and T. Lennartsson. 2000. Tolerance in plant ecology and evolution: toward a more unified theory of plant–herbivore interaction. Evol. Ecol. 14:283–287.

Karieva, P. 1997. Why worry about the maturing of a science? Ecoforum discussion, www.nceas.ucsb.edu.

Karr, J. R., and F. C. James. 1975. Ecomorphological configurations and convergent evolution. In Ecology and evolution of communities, ed. M. L. Cody and J. M. Diamond, 258–291. Cambridge: Harvard University Press.

Kaunzinger, C. M. K., and P. J. Morin. 1998. Productivity controls food-chain properties in microbial communities. Nature 395:495–497.

Kinzig, A. P., and S. W. Pacala. 2002. Successional biodiversity and ecosystem functioning. In The functional consequences of biodiversity: empirical progress and theoretical extensions, ed. A. P. Kinzig, S. W. Pacala, and D. Tilman, 175–212. Princeton: Princeton University Press.

Kinzig, A. P., S. W. Pacala, and D. Tilman, eds. 2002. The functional consequences of biodiversity: empirical progress and theoretical extensions. Princeton: Princeton University Press.

Kleijn, D., and J. M. Van Groenendael. 1999. The exploitation of heterogeneity by a clonal plant in habitats with contrasting productivity levels. J. Ecol. 87:873–884.

Knight, T. M. 2002. The effects of herbivory and pollen limitation on the population dynamics of *Trillium grandiflorum*. Ph.D. diss., University of Pittsburgh.

Kotler, B. P. 1984. Predation risk and the structure of desert rodent communities. Ecology 65:689–701.

Kotler, B. P., and J. S. Brown. 1999. Mechanisms of coexistence of optimal foragers as determinants of local abundances and distributions of desert granivores. J. Mammal. 80:361–374.

Kotler, B. P., J. S. Brown, S. R. X. Dall, S. Gresser, D. Ganey, and A. Bouskila. 2002. Foraging games between gerbils and their predators: temporal dynamics of resource depletion and apprehension in gerbils. Evol. Ecol. Res. 4:495–518.

Kotler, B. P., J. S. Brown, A. Oldfield, J. Thorson, and D. Cohen. 2001. Foraging substrate and escape substrate: patch use by three species of gerbils. Ecology 82:1781–1790.

Kraaijeveld, A. R., and H. C. J. Godfray. 1997. Trade-off between parasitoid resistance and larval competitive ability in *Drosophila melanogaster*. Nature 389:278–280.

Krebs, C. J. 2000. Ecology: the experimental analysis of distribution and abundance. 5th ed. Menlo Park, Calif.: Benjamin/Cummings.

Lack, D. 1940. Evolution of the Galapagos finches. Nature 146:324–327.

Lack, D. 1947. Darwin's finches. Cambridge: Cambridge University Press.

Laland, K. N., F. J. Odling-Smee, and M. W. Feldman. 1996. The evolutionary consequences of niche construction: a theoretical investigation using two-locus theory. J. Evol. Biol. 9:293–316.

Laland, K. N., F. J. Odling-Smee, and M. W. Feldman. 1999. Evolutionary consequences of niche construction and their implications for ecology. Proc. Nat. Acad. Sci. (USA) 96:10242–10247.

Lande, R. 1996. Statistics and partitioning of species diversity and similarity among multiple communities. Oikos 76:393–401.

Laska, M. S., and J. T. Wootton. 1998. Theoretical concepts and empirical approaches to measuring interaction strength. Ecology 79:461–476.

Law, R. 1999. Theoretical aspects of community assembly. In Advanced ecological theory: principles and applications, ed. J. McGlade, 141–171. Oxford: Blackwell.

Law, R., and R. D. Morton. 1993. Alternative permanent states of ecological communities. Ecology 74:1347–1361.

Law, R., and R. D. Morton. 1996. Permanence and the assembly of ecological communities. Ecology 77:762–775.

Lawler, S. P. 1993. Direct and indirect effects in microcosm communities of protists. Oecologia 93:184–190.

Lawler, S. P. 1998. Ecology in a bottle: using microcosms to test theory. In Experimental ecology: issues and perspectives, ed. W. J. Resetarits and J. Bernardo, 236–253. Oxford: Oxford University Press.

Lawler, S. P., and P. J. Morin. 1993. Food web architecture and population dynamics in laboratory microcosms of protists. Am. Nat. 141:675–686.

Lawton, J. H. 1999. Are there general laws in ecology? Oikos 84:177–192.

Lawton, J. H. 2000. Community ecology in a changing world. Excellence in Ecology, no. 11. Luhe, Germany: Ecology Institute.

Lawton, J. H., and V. K. Brown. 1993. Redundancy in ecosystems. In Biodiversity and ecosystem function, ed. E. D. Schulze and H. A. Mooney, 255–270. New York: Springer-Verlag.

Leibold, M. A. 1989. Resource edibility and the effects of predators and productivity on the outcome of trophic interactions. Am. Nat. 134:922–949.

Leibold, M. A. 1995. The niche concept revisited: mechanistic models and community context. Ecology 76:1371–1382.

Leibold, M. A. 1996. A graphical model of keystone predators in food webs: trophic regulation of abundance, incidence, and diversity patterns in communities. Am. Nat. 147:784–812.

Leibold, M. A. 1997. Do nutrient competition models predict nutrient availabilities in limnetic ecosystems? Oecologia 110:132–142.

Leibold, M. A. 1998. Similarity and local coexistence of species from regional biotas. Evol. Ecol. 12:95–110.

Leibold, M. A. 1999. Biodiversity and nutrient enrichment in pond plankton communities. Evol. Ecol. Res. 1:73–95.

Leibold, M. A., and A. J. Tessier. 1997. Habitat structure and the structure of lake plankton communities. In Ecology and evolution of freshwater animals, ed. B. Streit, T. Stadler, and C. J. Lively, 3–30. Basel: Birkhäuser.

Leibold, M. A., and G. M. Mikkelson. 2002. Coherence, species turnover, and boundary clumping: elements of meta-community structure. Oikos 97:237–250.

Leibold, M. A., and H. M. Wilbur. 1992. Interactions between food web structure and nutrients on pond organisms. Nature 360:341–343.

Leibold, M. A., J. M. Chase, J. B. Shurin, and A. Downing. 1997. Species turnover and the regulation of trophic structure. Annu. Rev. Ecol. Syst. 28:467–494.

Lenski, R. E. 1988. Experimental studies of pleiotropy and epistasis in *Escherichia coli*. Part 1. Variation in competitive fitness among mutants resistant to virus T4. Evolution 42:425–432.

Leon, J., and D. Tumpson. 1975. Competition between two species for complementary or substitutable resources. J. Theor. Biol. 50:185–201.

Levin, S. A. 1970. Community equilibria and stability, and an extension of the competitive exclusion principle. Am. Nat. 104:413–423.

Levin, S. A. 1974. Dispersion and population interactions. Am. Nat. 108:207–228.
Levin, S. A. 1999. Fragile dominion: complexity and the commons. Reading, Mass.: Perseus.
Levin, S. A., and R. T. Paine. 1974. Disturbance, patch formation, and community structure. Proc. Nat. Acad. Sci. (USA) 71:2744–2747.
Levine, J. M., and C. M. D'Antonio. 1999. Elton revisited: a review of evidence linking diversity and invasibility. Oikos 87:15–26.
Levine, J. M., and M. Rees. 2002. Coexistence and relative abundance in annual plant assemblages: the roles of competition and colonization. Am. Nat. 160:452–467.
Levins, R. 1968. Evolution in changing environments. Princeton: Princeton University Press.
Levins, R. 1979. Coexistence in a variable environment. Am. Nat. 114:765–783.
Levins, R., and D. Culver. 1971. Regional coexistence of species and competition between rare species. Proc. Nat. Acad. Sci. (USA) 68:1246–1248.
Levins, R., and R. Lewontin. 1980. Dialectics and reductionism in ecology. Synthese 43:47–78.
Lewin, R. 1983. Santa Rosalia was a goat. Science 221:636–639.
Liebig, J. von. 1840. Chemistry and its application to agriculture and physiology. London: Taylor and Walton.
Lindeman, R. L. 1942. The trophic dynamic aspect of ecology. Ecology 23:399–418.
Linnaeus, C. 1758. Systema Naturae: Creationis telluris est gloria Dei ex opere Naturae per Hominem solum.
Litchman, E., and C. A. Klausmeier. 2001. Competition of phytoplankton under fluctuating light. Am. Nat. 157:170–187.
Lively, C. M., W. N. Hazel, M. J. Schellenberger, and K. S. Michelson. 2000. Predator-induced defense: variation for inducibility in an intertidal barnacle. Ecology 81: 1240–1247.
Lomolino, M. V. 1996. Investigating causality of nestedness of insular communities: selective immigrations or extinctions? J. Biogeogr. 23:699–703.
Loreau, M. 1998. Biodiversity and ecosystem functioning: a mechanistic model. Proc. Nat. Acad. Sci. (USA) 95:5632–5636.
Loreau, M. 2000a. Are communities saturated? On the relationship between α, β, and γ diversity. Ecol. Letters 3:73–76.
Loreau, M. 2000b. Biodiversity and ecosystem functioning: recent theoretical advances. Oikos 91:3–17.
Loreau, M., and N. Mouquet. 1999. Immigration and the maintenance of local species diversity. Am. Nat. 154:427–440.
Loreau, M., S. Naeem, P. Inchausti, J. Bengtsson, J. P. Grim, A. Hector, D. U. Hooper, M. A. Huston, D. Raffaelli, B. Schmid, D. Tilman, and D. A. Wardle. 2001. Biodiversity and ecosystem functioning: current knowledge and future challenges. Science 294:804–808.
Losos, J. B., and D. Schluter. 2000. Analysis of an evolutionary species-area relationship. Nature 408:847–850.
Losos, J. B., S. Naeem, and R. K. Colwell. 1989. Hutchinsonian ratios and statistical power. Evolution 43:1820–1826.
Losos, J. B., T. R. Jackman, A. Larson, K. de Queiroz, and L. Rodriguez-Schettino. 1998. Contingency and determinism in replicated adaptive radiations of island lizards. Science 279:2115–2118.
Lotka, A. J. 1924. Elements of physical biology. Baltimore: Williams and Wilkins.
Lubchenco, J. 1978. Plant species diversity in a marine intertidal community: importance of herbivore food preference and algal competitive abilities. Am. Nat. 112: 23–39.

Luh, H. K., and S. L. Pimm. 1993. The assembly of ecological communities: a minimalist approach. J. Anim. Ecol. 62:749–765.
MacArthur, R. H. 1955. Fluctuations of animal populations, and a measure of community stability. Ecology 36:533–536.
MacArthur, R. H. 1958. Population ecology of some warblers of northeastern coniferous forests. Ecology 39:599–619.
MacArthur, R. H. 1965. Patterns of species diversity. Biol. Rev. 40:510–533.
MacArthur, R. H. 1969. The theory of the niche. In Population biology and evolution, ed. R. C. Lewontin, 159–176. Syracuse: Syracuse University Press.
MacArthur, R. H. 1972. Geographical ecology. Princeton: Princeton University Press.
MacArthur, R. H., and E. O. Wilson. 1967. The theory of island biogeography. Princeton: Princeton University Press.
MacArthur, R. H., and E. Pianka. 1966. On optimal use of a patchy environment. Am. Nat. 100:603–609.
MacArthur, R. H., and R. Levins. 1964. Competition, habitat selection, and character displacement in a patchy environment. Proc. Nat. Acad. Sci. (USA) 51:1207–1210.
MacArthur, R. H., and R. Levins. 1967. The limiting similarity, convergence, and divergence of coexisting species. Am. Nat. 101:377–385.
MacFadyen, A. 1957. Animal ecology: aims and methods. London: Pitman.
Mackey, R. L., and D. J. Currie. 2001. The diversity-disturbance relationship: is it generally strong and peaked? Ecology 82:3479–3492.
MacNally, R. C. 1995. Ecological versatility and community ecology. Cambridge: Cambridge University Press.
Maguire, B. 1973. Niche response structure and the analytical potentials of its relationship to habitat. Am. Nat. 107:213–243.
Margalef, R. 1974. Diversity, stability, and maturity in natural ecosystems. Proc. 1st Int. Cong. Ecol. 66.
Maurer, B. A. 1999. Untangling ecological complexity. Chicago: University of Chicago Press.
May, R. M. 1974a. Biological populations with nonoverlapping generations: stable points, stable cycles, and chaos. Science 186:645–647.
May, R. M. 1974b. Stability and complexity in model ecosystems. 2d ed. Princeton: Princeton University Press.
May, R. M., and R. H. MacArthur. 1972. Niche overlap as a function of environmental variability. Proc. Nat. Acad. Sci. (USA) 69:1109–1113.
Maynard Smith, J. 1976. Evolution and the theory of games. Cambridge: Cambridge University Press.
Mayr, E., and W. B. Provine, eds. 1998. The evolutionary synthesis: perspectives on the unification of biology. 2d ed. Cambridge: Harvard University Press.
McCann, K. S. 2000. The diversity-stability debate. Nature 405:228–233.
McCann, K. S., A. Hastings, and G. R. Huxel. 1998. Weak trophic interactions and the balance of nature. Nature 395:794–798.
McCune, B., and T. F. H. Allen. 1985. Will similar forests develop on similar sites? Can. J. Bot. 63:367–376.
McGrady-Steed, J., P. M. Harris, and P. J. Morin. 1997. Biodiversity regulates ecosystem predictability. Nature 390:162–165.
McIntosh, R. P. 1985. The background of ecology. Cambridge: Cambridge University Press.
McNaughton, S. J. 1977. Diversity and stability of ecological communities: a comment on the role of empiricism in ecology. Am. Nat. 111:515–525.
McPeek, M. A. 1996. Trade-offs, food web structure, and the coexistence of habitat specialists and generalists. Am. Nat. 148: S124-S138.

McPeek, M. A. 1998. The consequences of changing the top predator in a food web: a comparative experimental approach. Ecol. Monogr. 68:1–23.
McPeek, M. A., and B. L. Peckarsky. 1998. Life histories and the strengths of species interactions: combining mortality, growth, and fecundity effects. Ecology 79:867–879.
McPeek, M. A., and J. M. Brown. 2000. Building a regional species pool: Diversification of the Enallagma damselflies in eastern North America. Ecology 81:904–920.
McPeek, M. A., and T. E. Miller. 1996. Evolutionary biology and community ecology. Ecology 77:1319–1320.
Merriam-Webster's Collegiate Dictionary. 10th ed. 2001. Springfield, Mass.: Merriam-Webster.
Merriam-Webster's Collegiate Dictionary. 5th ed. 1959. Springfield, Mass.: Merriam-Webster.
Miller, R. S. 1967. Patterns and process in competition. Adv. Ecol. Res. 4:1–74.
Miller, T. E., and J. Travis. 1996. The evolutionary role of indirect effects in communities. Ecology 77:1329–1335.
Miller, T. E., and W. C. Kerfoot. 1987. Redefining indirect effects. In Predation: direct and indirect effects on aquatic communities, ed. W. C. Kerfoot and A. Sih, 33–37. Hanover, N.H.: Hanover.
Miller, T. E., J. M. Kneitel, and J. H. Burns. 2002. Effect of community structure on invasion success and rate. Ecology 83:898–905.
Mitchley, J., and P. J. Grubb. 1986. Control of relative abundance of perennials in chalk grassland in southern England. Part 1. Constance of rank order and results of pot- and field-experiments on the role of interference. J. Ecol. 74:1139–1166.
Mittelbach, G. G., C. F. Steinter, K. L. Gross, H. L. Reynolds, S. M. Scheiner, R. B. Waide, M. R. Willig, and S. I. Dodson. 2001. What is the observed relationship between species richness and productivity? Ecology 82:2381–2396.
Moen, J., and L. Oksanen. 1992. Ecosystem trends. Nature 353:510.
Moen, J., H. Garolfjell, L. Oksanen, L. Ericson, and P. Ekerholm. 1996. Grazing by food-limited microtine rodents on a productivity experimental plant community: does the green desert exist? Oikos 68:401–413.
Morin, P. J. 1999. Community ecology. Cambridge: Cambridge University Press.
Morton, R. D., R. Law, S. L. Pimm, and J. A. Drake. 1996. On models for assembling ecological communities. Oikos 75:493–499.
Mouquet, N., and M. Loreau. 2002. Coexistence in metacommunities: the regional similarity hypothesis. Am. Nat. 159:420–426.
Naeem, S. 1998. Species redundancy and ecosystem reliability. Conserv. Biol. 12:39–45.
Naeem, S., and S. B. Li. 1997. Biodiversity enhances ecosystem reliability. Nature 390: 507–509.
Naeem, S., F. S. Chapin, R. Costanza, P. R. Ehrlich, F. Golley, D. U. Hooper, J. H. Lawton, R. V. O'Neill, H. A. Mooney, O. E. Sala, A. J. Symstad, and D. Tilman. 2000. Biodiversity and ecosystem functioning: maintaining natural life support processes. Issues in Ecology 4. Washington: Ecological Society of America.
Neill, W. E. 1975. Experimental studies of microcrustacean competition, community competition, and efficiency of resource utilization. Ecology 56:809–826.
Neubert, M. G., and H. Caswell. 1997. Alternatives to resilience for measuring the responses of ecological systems to perturbations. Ecology 78:653–665.
Neuhauser, C., D. Andow, G. Heimpel, G. May, R. Shaw, and S. Wagenius. 2003. Community genetics: expanding the synthesis of ecology and genetics. Ecology 84: (in press).
Newman, R. A. 1998. Ecological constraints on amphibian metamorphosis: interactions of temperature and larval density with responses to changing food level. Oecologia 115:9–16.

Odling-Smee, F. J., K. N. Laland, and M. W. Feldman. 1996. Niche construction. Am. Nat. 147:641–648.
Odum, E. P. 1951. Fundamentals of ecology. Philadelphia: Saunders.
Odum, E. P. 1971. Fundamentals of ecology. 3d ed. Philadelphia: Saunders.
Ohta, T. 1992. The nearly neutral theory of molecular evolution. Annu. Rev. Ecol. Syst. 23:263–286.
Oksanen, L., and T. Oksanen. 2000. The logic and realism of the hypothesis of exploitation ecosystems. Am. Nat. 155:703–723.
Oksanen, L., S. Fretwell, J. Arruda, and P. Niemala. 1981. Exploitation ecosystems in gradients of primary productivity. Am. Nat. 118:240–261.
Osenberg, C. W. 1989. Resource limitation, competition, and the influence of life history in a freshwater snail community. Oecologia 79:512–519.
Osenberg, C. W., G. G. Mittelbach, and P. C. Wainwright. 1992. Two-stage life histories in fish: the interaction between juvenile competition and adult performance. Ecology 73:255–267.
Pacala, S. W., and D. Tilman. 1994. Limiting similarity in mechanistic and spatial models of plant competition in heterogeneous environments. Am. Nat. 143:222–257.
Pacala, S. W., and M. Rees. 1998. Models suggesting field experiments to test two hypotheses explaining successional diversity. Am. Nat. 152:729–737.
Paine, R. T. 1966. Food web complexity and species diversity. Am. Nat. 100:65–75.
Paine, R. T. 2002. Trophic control of production in a rocky intertidal community. Science 296:736–739.
Paine, R. T., and S. A. Levin. 1981. Intertidal landscapes: disturbance and the dynamics of pattern. Ecol. Monogr. 51:145–178.
Paine, R. T., J. C. Castillo, and J. Cancino. 1985. Perturbation and recovery patterns of starfish-dominated intertidal assemblages in Chile, New Zealand, and Washington State. Am. Nat. 125:679–691.
Pake, C. E., and D. L. Venable. 1996. Seed banks in desert annuals: implications for persistence and coexistence in variable environments. Ecology 77:1427–1435.
Pascual, M., and S. A. Levin. 1999. From individuals to population densities: searching for the intermediate scale of nontrivial determinism. Ecology 80:2225–2236.
Pastor, J., and Y. Cohen. 1997. Herbivores, the functional diversity of plant species, and the cycling of nutrients in ecosystems. Theor. Pop. Biol. 51:165–179.
Patrick, R., ed. 1983. Diversity. Stroudsburg, Pa.: Hutchinson Ross.
Patterson, B. D. 1987. The principle of nested subsets and its implications for biological conservation. Conserv. Biol. 1:323–334.
Persson, A., L. A. Hansson, C. Bronmark, P. Lundberg, L. B. Pettersson, L. Greenberg, P. A. Nilsson, P. Nystrom, P. Romare, and L. Tranvik. 2001. Effects of enrichment on simple aquatic food webs. Am. Nat. 157:654–669.
Persson, L., P. Bystrom, E. Wahlstrom, J. Andersson, and J. Hjelm. 1999. Interactions among size-structured populations in a whole-lake experiment: size- and scale-dependent processes. Oikos 87:139–156.
Petchey, O. L. 2000. Species diversity, species extinction, and ecosystem function. Am. Nat. 155:696–702.
Petchey, O. L., P. T. McPhearson, T. M. Casey, and P. J. Morin. 1999. Environmental warming alters food-web structure and ecosystem function. Nature 402:69–72.
Petraitis, P. S. 1989. The representation of niche breadth and overlap on Tilman's consumer-resource graphs. Oikos 56:289–292.
Petraitis, P. S., and R. E. Latham. 1999. The importance of scale in testing the origins of alternative community states. Ecology 80:429–442.
Petraitis, P. S., and S. R. Dudgeon. 1999. Experimental evidence for the origin of alternative communities on rocky intertidal shores. Oikos 84:239–245.

Petraitis, P. S., R. E. Latham, and R. A. Niesenbaum. 1989. The maintenance of species diversity by disturbance. Quart. Rev. Biol. 64:393–418.

Pianka, E. R. 1973. The structure of lizard communities. Annu. Rev. Ecol. Syst. 4:53–74.

Pianka, E. R. 1995. Evolutionary ecology. 5th ed. New York: HarperCollins.

Pianka, E. R. 1999. Putting communities together. Trends Ecol. Evol. 14:501–502.

Pickett, S. T. A., and P. S. White, eds. 1985. The ecology of natural disturbance and patch dynamics. San Diego: Academic Press.

Pickett, S. T. A., R. S. Ostfeld, and M. Shachak, eds. 1997. The ecological basis of conservation: heterogeneity, ecosystems, and biodiversity. New York: Chapman and Hall.

Pimm, S. L. 1991. The balance of nature? ecological issues in the conservation of species and communities. Chicago: University of Chicago Press.

Pitman, N. C. A., J. W. Terborgh, M. R. Silman, P. Núñez, D. A. Neill, C. E. Ceron, W. A. Palacios, and M. Aulestia. 2001. Dominance and distribution of tree species in upper Amazonian terra firme forests. Ecology 82:2101–2117.

Polis, G. A. 1999. Why are parts of the world green? multiple factors control productivity and the distribution of biomass. Oikos 86:3–15.

Polis, G. A., A. L. W. Sears, G. R. Huxel, D. R. Strong, and J. Maron. 2000. When is a trophic cascade a trophic cascade? Trends Ecol. Evol. 15:473–475.

Polis, G. A., and D. R. Strong. 1996. Food web complexity and community dynamics. Am. Nat. 147:813–846.

Popper, K. R. 1963. Conjectures and refutations: the growth of scientific knowledge. New York: Basic Books.

Power, M. E. 1992. Top-down and bottom-up forces in food webs: do plants have primacy? Ecology 73:733–746.

Power, M. E., D. Tilman, J. A. Estes, B. A. Menge, W. J. Bond, L. S. Mills, G. Daily, J. C. Castilla, J. Lubchenco, and R. T. Paine. 1996. Challenges in the quest for keystones. Bioscience 46:609–620.

Pulliam, H. R. 1988. Sources, sinks and population regulation. Am. Nat. 132:652–661.

Pulliam, H. R. 2000. On the relationship between niche and distribution. Ecol. Letters 3:349–361.

Quammen, D. 1996. The song of the dodo: island biogeography in an age of extinctions. New York: Scribner.

Rainey, P. B., and M. Travisano. 1998. Adaptive radiation in a heterogeneous environment. Nature 394:69–72.

Rausher, M. D. 2001. Co-evolution and plant resistance to natural enemies. Nature 411:857–864.

Reaka-Kudla, M. L., D. E. Wilson, and E. O. Wilson, eds. 1997. Biodiversity II: understanding and protecting our biological resources. Washington: Joseph Henry.

Real, L. A., and S. A. Levin. 1991. Theoretical advances: the role of theory in the rise of modern ecology. In Foundations of ecology: classic papers with commentaries, ed. L. A. Real and J. H. Brown, 177–191. Chicago: University of Chicago Press.

Real, L. A., and T. Caraco. 1986. Risk and foraging in stochastic environments. Annu. Rev. Ecol. Syst. 17:371–390.

Rees, M., and J. Bergelson. 1997. Asymmetric light competition and founder control in plant communities. J. Theor. Biol. 184:353–358.

Resetarits, W. J., and J. Bernardo, J., eds. 1998. Experimental ecology: issues and perspectives. Oxford: Oxford University Press.

Reynolds, H. R., and S. W. Pacala. 1993. An analytical treatment of root-to-shoot ratio and plant competition for soil nutrient and light. Am. Nat. 141:51–70.

Ricklefs, R. E. 1987. Community diversity: relative roles of local and regional processes. Science 235:167–171.

Ricklefs, R. E., and D. Schluter, eds. 1993. Species diversity in ecological communities: historical and geographical perspectives. Chicago: University of Chicago Press.

Ricklefs, R. E., and G. L. Miller. 2000. Ecology. 4th ed. New York: Freeman.

Ritchie, M. E., and D. Tilman. 1995. Responses of legumes to herbivores and nutrients during succession on a nitrogen-poor soil. Ecology 76:2648–2655.

Ritchie, M. E., and H. Olff. 1999. Spatial scaling laws yield a synthetic theory of biodiversity. Nature 400:557–560.

Ritchie, M. E., D. Tilman, and J. M. H. Knops. 1998. Herbivore effects on plant and nitrogen dynamics in oak savanna. Ecology 79:165–177.

Robinson, J. V., and J. E. Dickerson. 1987. Does invasion sequence affect community structure? Ecology 68:587–595.

Robinson, J. V., and M. A. Edgemon. 1988. An experimental evaluation of the effect of invasion history on community structure. Ecology 69:1410–1417.

Root, R. B. 1967. The niche exploitation pattern of the blue-grey gnatcatcher. Ecol. Monogr. 37:317–349.

Rosenzweig, M. L. 1971. The paradox of enrichment: destabilization of exploitation ecosystems in ecological time. Science 171:385–387.

Rosenzweig, M. L. 1981. A theory of habitat selection. Ecology 62:327–335.

Rosenzweig, M. L. 1995. Species diversity in space and time. Cambridge: Cambridge University Press.

Rosenzweig, M. L. 1999. Species diversity. In Advanced ecological theory: principles and applications, ed. J. McGlade, 249–281. Oxford: Blackwell.

Rosenzweig, M. L., and Z. Abramsky. 1993. How are diversity and productivity related? In Species diversity in ecological communities: historical and geographical perspectives, ed. R. E. Ricklefs and D. Schluter, 52–65. Chicago: University of Chicago Press.

Rosenzweig, M. L., and Z. Abramsky. 1997. Two gerbils of the Negev: A long-term investigation of optimal habitat selection and its consequences. Evol. Ecol. 11:733–756.

Rothaupt, K. O. 1988. Mechanistic resource competition theory applied to laboratory experiments with zooplankton. Nature 333:660–662.

Roughgarden, J. 1972. Evolution of niche width. Am. Nat. 106:683–718.

Roughgarden, J. 1976. Resource partitioning among competing species: a coevolutionary approach. Theor. Pop. Biol. 9:388–424.

Ryel, R. J., and M. M. Caldwell. 1998. Nutrient acquisition from soils with patchy nutrient distributions as assessed with simulation models. Ecology 79:2735–2744.

Salisbury, E. J. 1929. The biological equipment of species in relation to competition. J. Ecol. 17:197–222.

Samuels, C. L., and J. A. Drake. 1997. Divergent perspectives on community convergence. Trends Ecol. Evol. 12:427–432.

Schaffer, W. M. 1981. Ecological abstraction: the consequences of reduced dimensions in ecological models. Ecol. Monogr. 51:383–401.

Schindler, D. W. 1990. Experimental perturbations of whole lakes as tests of hypotheses concerning ecosystem structure and function. Oikos 57:25–41.

Schlichting, C. D., and M. Pigliucci. 1998. Phenotypic evolution: a reaction norm perspective. Sunderland, Mass.: Sinauer.

Schluter, D. 1996. Ecological causes of adaptive radiation. Am. Nat. 148, suppl.: S40–S64.

Schluter, D. 2000. The ecology of adaptive radiation. Oxford: Oxford University Press.

Schluter, D. 2001. Ecology and the origin of species. Trends Ecol. Evol. 16:372–380.

Schmida, A., and S. Ellner. 1984. Coexistence of plant species with similar niches. Vegetatio 58:29–55.

Schmitz, O. J. 1997. Press perturbations and the predictability of ecological interactions in a food web. Ecology 78:55–69.

Schoener, T. W. 1965. The evolution of bill size differences among sympatric congeneric species of birds. Evolution 19:189–203.

Schoener, T. W. 1968. The *Anolis* lizards of Bimini: resource partitioning in a complex fauna. Ecology 49:704–726.

Schoener, T. W. 1974a. Resource partitioning in ecological communities. Science 185: 27–39.

Schoener, T. W. 1974b. Some methods for calculating competition coefficients from resource-utilization spectra. Am. Nat. 108:332–340.

Schoener, T. W. 1983. Field experiments on interspecific competition. Am. Nat. 122: 155–174.

Schoener, T. W. 1984. Size differences among sympatric, bird-eating hawks: a worldwide survey. In Ecological communities: conceptual issues and the evidence, ed. D. R. Strong, D. S. Simberloff, L. G. Abele, and A. B. Thistle, 254–281. Princeton: Princeton University Press.

Schoener, T. W. 1989. The ecological niche. In Ecological concepts, ed. J. M. Cherrett, 79–113. Oxford: Blackwell.

Schwartz, M. W., and J. D. Hoeksema. 1998. Specialization and resource trade: biological markets as a model of mutualisms. Ecology 79:1029–1038.

Shea, K., and P. L. Chesson. 2002. Community ecology theory as a framework for biological invasions. Trends Ecol. Evol. 17:170–176.

Shurin, J. B. 2000. Dispersal limitation, invasion resistance, and the structure of pond zooplankton communities. Ecology 81:3074–3086.

Shurin, J. B. 2001. Interactive effects of predation and dispersal on zooplankton communities. Ecology 82:3404–3416.

Sih, A., P. Crowley, M. McPeek, J. Petranka, and K. Strohmeier. 1985. Predation, competition, and prey communities: a review of field experiments. Annu. Rev. Ecol. Syst. 16:269–311.

Simberloff, D. S. 1970. Taxonomic diversity of island biotas. Evolution 24:23–37.

Simberloff, D. S. 1978. Using island biogeographic distributions to determine if colonization is stochastic. Am. Nat. 112:713–726.

Simberloff, D. S., and W. Boecklen. 1981. Santa Rosalia reconsidered: size ratios and competition. Evolution 35:1206–1228.

Simms, E. L. 1992. Costs of plant resistance to herbivory. In Plant resistance to herbivores and pathogens: ecology, evolution, and genetics, ed. R. S. Fritz and E. L. Simms, 392–425. Chicago: University of Chicago Press.

Simpson, G. G. 1953. The major features of evolution. New York: Columbia University Press.

Sinclair, A. R. E., C. J. Krebs, J. M. Fryxell, R. Turkington, S. Boutin, R. Boonstra, P. Seccombe-Hett, P. Lundberg, and L. Oksanen. 2000. Testing hypotheses of trophic level interactions: a boreal forest ecosystem. Oikos 89:313–328.

Skelly, D. K. 1995. A behavioral trade-off and its consequences for the distribution of *Pseudacris* treefrog larvae. Ecology 76:150–164.

Skelly, D. K., and E. E. Werner. 1990. Behavioral and life-historical responses of larval American toads to an odonate predator. Ecology 71:2313–2322.

Smith, R. L., and T. Smith. 2000. Ecology and field biology. Boston: Addison-Wesley.

Smith, V. H. 1983. Low nitrogen to phosphorus ratios favor dominance by blue-green algae in lake phytoplankton. Science 221:669–671.

Sommer, U. 1990. Phytoplankton nutrient competition: from laboratory to lake. In

Perspectives on plant competition, ed. J. B. Grace and D. Tilman, 193–213. San Diego: Academic Press.

Sommer, U. 1991. Convergent succession of phytoplankton in microcosms with different inoculum species composition. Oecologia 87:171–179.

Sousa, W. P. 1979. Disturbance in marine intertidal boulder fields: the nonequilibrium maintenance of species diversity. Ecology 60:1225–1239.

Sousa, W. P. 1984. The role of disturbance in natural communities. Annu. Rev. Ecol. Syst. 15:353–391.

Stauffer, Robert C. 1975. Charles Darwin's natural selection: being the second part of his big species book written from 1856 to 1858. Cambridge: Cambridge University Press.

Stearns, S. C. 1992. The evolution of life-histories. Oxford: Oxford University Press.

Steiner, C. F. 2001. The effects of prey heterogeneity and consumer identity on the limitation of trophic-level biomass. Ecology 82:2495–2506.

Stone, L., T. Dayan, and D. S. Simberloff. 2000. On desert rodents, favored states, and unresolved issues: scaling up and down regional assemblages and local communities. Am. Nat. 156:322–328.

Stowe, K. A., R. J. Marquis, C. G. Hochwender, and E. L. Simms. 2000. The evolutionary ecology of tolerance to consumer damage. Annu. Rev. Ecol. Syst. 31:565–595.

Strauss, S. Y., and A. A. Agrawal. 1999. The ecology and evolution of plant tolerance to herbivory. Trends Ecol. Evol. 14:179–185.

Strong, D. R. 1992. Are trophic cascades all wet? Differentiation and donor-control in speciose ecosystems. Ecology 73:747–754.

Strong, D. R., L. A. Szyska, and D. S. Simberloff. 1979. Tests of community-wide character displacement against null hypotheses. Evolution 33:897–913.

Tansley, A. G. 1917. On competition between *Galium saxatile* L. (*G. hercynicum* Weig.) and *Galium sylvestre* Poll. (*G. asperum* Schreb.) on different types of soil. J. Ecol. 5: 173–179.

Tansley, A. G. 1935. The use and abuse of vegetation concepts and terms. Ecology 16: 284–307.

Taper, M. L., and T. J. Case. 1985. Quantitative genetic models for the coevolution of character displacement. Ecology 66:355–371.

Terborgh, J., R. B. Foster, and P. Núñez. 1996. Tropical tree communities: A test of the nonequilibrium hypothesis. Ecology 77:561–567.

Tessier, A. J., and P. Woodruff. 2002. Cryptic trophic cascade along a gradient of lake size. Ecology 83:1263–1270.

Tessier, A. J., M. A. Leibold, and J. Tsao. 2000. A fundamental trade-off in resource exploitation by *Daphnia* and consequences to plankton communities. Ecology 81: 826–841.

Thompson, J. 1994. The coevolutionary process. Chicago: University of Chicago Press.

Thompson, J., O. J. Reichman, P. J. Morin, G. A. Polis, M. E. Power, R. W. Sterner, C. A. Couch, L. Gough, R. Holt, D. U. Hooper, F. Keesing, C. R. Lovell, B. T. Milne, M. C. Molles, D. W. Roberts, and S. Y. Strauss. 2001. Frontiers of ecology. Bioscience 51:15–24.

Tiffin, P. 2000. Are tolerance, avoidance, and antibiosis evolutionarily and ecologically equivalent responses of plants to herbivores? Am. Nat. 155:128–138.

Tilman, D. 1976. Ecological competition between algae: experimental confirmation of resource-based competition theory. Science 192:463–465.

Tilman, D. 1977. Resource competition between planktonic algae: an experimental and theoretical approach. Ecology 58:338–348.

Tilman, D. 1980. Resource: a graphical-mechanistic approach to competition and predation. Am. Nat. 116:362–393.

Tilman, D. 1982. Resource competition and community structure. Princeton: Princeton University Press.

Tilman, D. 1985. The resource-ratio hypothesis of plant succession. Am. Nat. 125: 827–852.

Tilman, D. 1988. Plant strategies and the dynamics and structure of plant communities. Princeton: Princeton University Press.

Tilman, D. 1990. Constraints and trade-offs: toward a predictive theory of competition and succession. Oikos 58:3–15.

Tilman, D. 1994. Competition and biodiversity in spatially structured habitats. Ecology 75:2–16.

Tilman, D. 1996. Biodiversity: population versus ecosystem stability. Ecology 77:350–363.

Tilman, D. 1997. Community invasibility, recruitment limitation, and grassland biodiversity. Ecology 78:81–92.

Tilman, D. 1999. The ecological consequences of changes in biodiversity: a search for general principles. Ecology 80:1455–1474.

Tilman, D., and D. Wedin. 1991a. Dynamics of nitrogen competition between successional grasses. Ecology 72:1038–1049.

Tilman, D., and D. Wedin. 1991b. Plant traits and resource reduction for five grasses growing on a nitrogen gradient. Ecology 72:685–700.

Tilman, D., and J. A. Downing. 1994. Biodiversity and stability in grasslands. Nature 367:363–365.

Tilman, D., and P. Karieva, eds. 1997. Spatial ecology: the role of space in population dynamics and interspecific interactions. Princeton: Princeton University Press.

Tilman, D., and R. W. Sterner. 1984. Invasions of equilibria: tests of resource competition using two species of algae. Oecologia 61:197–200.

Tilman, D., and S. Pacala. 1993. The maintenance of species richness in plant communities. In Species diversity in ecological communities: historical and geographical perspectives, ed. R. E. Ricklefs and D. Schluter, 13–25. Chicago: University of Chicago Press.

Tilman, D., C. L. Lehman, and C. E. Bristow. 1998. Diversity-stability relationships: statistical inevitability or ecological consequence? Am. Nat. 151:277–282.

Tilman, D., C. L. Lehman, and K. T. Thomson. 1997. Plant diversity and ecosystem productivity: theoretical considerations. Proc. Nat. Acad. Sci. (USA) 94:1857–1861.

Tilman, D., D. Wedin, and J. Knops. 1996. Productivity and sustainability influenced by biodiversity in grassland ecosystems. Nature 379:718–720.

Tilman, D., J. Knops, D. Wedin, P. Reich, M. Ritchie, and E. Siemann. 1997. The influence of functional diversity and composition on ecosystem processes. Science 277:1300–1302.

Tilman, D., M. Mattson, and S. Langer. 1981. Competition and nutrient kinetics along a temperature gradient: an experimental test of a mechanistic approach to niche theory. Limnol. Oceanogr. 26:1020–1033.

Tilman, D., P. B. Reich, J. Knops, D. Wedin, T. Mielke, and C. Lehman. 2001. Diversity and productivity in a long-term grassland experiment. Science 294:843–845.

Tilman, D., R. Kiesling, R. Sterner, S. Kilham, and F. A. Johnson. 1986. Green, bluegreen, and diatom algae: taxonomic differences in competitive ability for phosphorus, silicon, and nitrogen. Arch. Hydrobiol. 106:473–485.

Tilman, D., S. S. Kilham, and P. Kilham. 1976. Morphometric changes in *Asterionella formosa* colonies under phosphate and silicate limitation. Limnol. Oceanogr. 21:883–886.

Tollrian, R., and C. D. Harvell, eds. 1998. The ecology and evolution of inducible defenses. Princeton: Princeton University Press.
Turelli, M. 1978. Does environmental variability limit niche overlap? Proc. Nat. Acad. Sci. (USA) 75:5085–5089.
Turelli, M. 1981. Niche overlap and invasion of competitors in random environments. Part 1. Models without demographic stochasticity. Theor. Pop. Biol. 20:1–56.
Udvardy, M. F. 1959. Notes on the ecological concepts of habitat, biotope, and niche. Ecology 40:725–728.
Van Buskirk, J. 2000. The costs of an inducible defense in anuran larvae. Ecology 81: 2813–2821.
Van Valen, L. M. 1965. Morphological variation and the width of the ecological niche. Am. Nat. 99:377–390.
Van Valen, L. M. 1973. A new evolutionary law. Evol. Theory 1:1–30.
Vance, R. R. 1978. Predation and resource partitioning in one-predator-two-prey model communities. Am. Nat. 112:797–813.
Vandermeer, J. H. 1972. Niche theory. Annu. Rev. Ecol. Syst. 3:107–132.
Vandermeer, J. H. 1975. Interspecific competition: a new approach to the classical theory. Science 188:253–255.
Vandermeer, J. H. 1992. The ecology of intercropping. Cambridge: Cambridge University Press.
Vanni, M. J., A. S. Flecker, J. M. Hood, and J. L. Headworth. 2002. Stoichiometry of nutrient recycling by vertebrates in a tropical stream: linking species identity and ecosystem processes. Ecol. Letters 5:285–293.
Vanni, M. J., C. D. Layne, and S. E. Arnott. 1997. "Top-down" trophic interactions in lakes: effects of fish on nutrient dynamics. Ecology 78:1–20.
Vermeij, G. J. 1987. Evolution and escalation: an ecological history of life. Princeton: Princeton University Press.
Vermeij, G. J. 1991. When biotas meet: understanding biotic interchange. Science 253: 1099–1104.
Vincent, T. L. S., D. Scheel, J. S. Brown, and T. L. Vincent. 1996. Trade-offs and coexistence in consumer-resource models: it all depends on what and where you eat. Am. Nat. 148:1038–1058.
Vincent, T. L., and J. S. Brown. 1988. The evolution of ESS theory. Annu. Rev. Ecol. Syst. 19:423–443.
Vincent, T. L., Y. Cohen, and J. S. Brown. 1993. Evolution via strategy dynamics. Theor. Pop. Biol. 44:149–176.
Volterra, V. 1926. Fluctuations in the abundance of a species considered mathematically. Nature 118:558–560.
Waide, R. B., M. R. Willig, C. F. Steiner, G. G. Mittelbach, L. Gough, S. I. Dodson, G. P. Juday, and R. Parameter. 1999. The relationship between primary productivity and species richness. Annu. Rev. Ecol. Syst. 30:257–300.
Walker, B. 1992. Conserving biological diversity through ecosystem resilience. Conserv. Biol. 9:747–752.
Wallace, A. R. 1876. The geographical distribution of animals. New York: Hafner.
Wang, Z.-L., F.-Z. Wang, S. Chen, and M.-Y. Zhu. 2002. Competition and coexistence in regional habitats. Am. Nat. 159:498–508.
Wardle, D. A. 1999. Is "sampling effect" a problem for experiments investigating biodiversity-ecosystem function relationships? Oikos 87:403–407.
Wardle, D. A., M. A. Huston, J. P. Grime, F. Berandse, E. Garnier, W. K. Lauenroth, H. Setala, and S. D. Wilson. 2000. Biodiversity and ecosystem function: an issue in ecology. Bull. Ecol. Soc. Am. 81:235–240.

Warner, R. R., and P. L. Chesson. 1985. Coexistence mediated by recruitment fluctuations: a field guide to the storage effect. Am. Nat. 125:769–787.

Watson, S., E. McCauley, and J. A. Downing. 1992. Sigmoid relationships between phosphorus, algal biomass, and algal community structure. Can. J. Fish. Aquat. Sci. 49:2605–2610.

Webb, C. O. 2000. Exploring the phylogenetic structure of ecological communities: an example for rainforest trees. Am. Nat. 156:145–155.

Wedin, D., and D. Tilman. 1993. Competition among grasses along a nitrogen gradient: initial conditions and mechanisms of competition. Ecol. Monogr. 63:199–229.

Weiher, E., and P. Keddy. 1995. The assembly of experimental wetland plant communities. Oikos 73:323–335.

Weiher, E., and P. Keddy. 1999. Assembly rules as general constraints on community composition. In Ecological assembly rules: perspectives, advances, retreats, ed. E. Weiher and P. A. Keddy, 251–271. Cambridge: Cambridge University Press.

Wellborn, G. A., D. K. Skelly, and E. E. Werner. 1996. Mechanisms creating community structure across a freshwater habitat gradient. Annu. Rev. Ecol. Syst. 27:337–363.

Werner, E. E. 1986. Amphibian metamorphosis: growth rate, predation risk, and the optimal size at transformation. Am. Nat. 128:319–341.

Werner, E. E., and D. J. Hall. 1976. Niche shift in sunfishes: experimental evidence and significance. Science 191:404–406.

Werner, E. E., and D. J. Hall. 1977. Competition and habitat shift in two sunfishes (Centrarchidae). Ecology 58:869–876.

Werner, E. E., and D. J. Hall. 1979. Foraging efficiency and habitat switching in competing sunfishes. Ecology 60:256–264.

Werner, E. E., and D. J. Hall. 1988. Ontogenetic habitat shifts in bluegill: the foraging rate–predation risk trade-off. Ecology 69:1352–1366.

Werner, E. E., and J. F. Gilliam. 1984. The ontogenetic niche and species interactions in size-structured populations. Annu. Rev. Ecol. Syst. 15:393–425.

Werner, E. E., J. F. Gilliam, D. J. Hall, and G. G. Mittelbach. 1983. An experimental test of the effects of predation risk on habitat use in fish. Ecology 64:1540–1548.

Whitlock, M. C. 1996. The red queen beats the jack-of-all-trades: the limitations on the evolution of phenotypic plasticity and niche breadth. Am. Nat. 148, suppl.: S65–S77.

Whittaker, R. H. 1956. Vegetation of the Great Smoky Mountains. Ecol. Monogr. 26: 1–80.

Whittaker, R. H. 1962. Classification of natural communities. Bot. Rev. 28:1–239.

Whittaker, R. H. 1967. Gradient analysis of vegetation. Biol. Rev. 42:207–264.

Whittaker, R. H. 1972. Evolution and the measurement of species diversity. Taxon 21: 213–251.

Whittaker, R. H., and S. A. Levin, eds. 1975. Niche: theory and application. Stroudsburg: Wiley.

Whittaker, R. H., S. A. Levin, and R. B. Root. 1973. Niche, habitat, and ecotope. Am. Nat. 107:3212–338.

Wilbur, H. M., and J. P. Collins. 1973. Ecological aspects of amphibian metamorphosis. Science 182:1305–1314.

Williams, C. B. 1964. Patterns in the balance of nature and related problems of quantitative biology. London: Academic Press.

Williamson, M. H. 1972. The analysis of biological populations. London: Edward Arnold.

Williamson, M. H. 1996. Biological invasions. London: Chapman and Hall.

Wilson, J. B., J. B. Steel, M. E. Dood, B. J. Anderson, I. Ullmann, and P. Bannister.

2000. A test of community reassembly using the exotic communities of New Zealand roadsides in comparison to British roadsides. J. Ecol. 88:757–764.

Wilson, S. D., and D. Tilman. 1993. Plant competition and resource availability in response to disturbance and fertilization. Ecology 74:599–611.

Wilson, S. D., and D. Tilman. 1995. Competitive responses of eight old-field plant species in four environments. Ecology 76:1169–1180.

Wilson, W. G., C. W. Osenberg, R. J. Schmitt, and R. M. Nisbet. 1999. Complementary foraging behaviors allow coexistence of two consumers. Ecology 80:2358–2372.

Wootton, J. T. 1994. The nature and consequences of indirect effects in ecological communities. Annu. Rev. Ecol. Syst. 25:443–466.

Wootton, J. T. 1997. Estimates and tests of per capita interaction strength: diet, abundance, and impact of intertidally foraging birds. Ecol. Monogr. 67:45–64.

Wootton, J. T. 1998. Effects of disturbance on species diversity: a multi-trophic perspective. Am. Nat. 152:803–825.

Worthen, W. B. 1996. Community composition and nested-subset analyses: basic descriptors for community ecology. Oikos 76:417–426.

Wright, D. H., B. D. Patterson, G. M. Mikkelson, A. Cutler, and W. Atmar. 1998. A comparative analysis of nested subset patterns of species composition. Oecologia 113:1–20.

Wright, S. 1931. Evolution in Mendelian populations. Genetics 16:97–159.

Yachi, S., and M. Loreau. 1999. Biodiversity and ecosystem productivity in a fluctuating environment: the insurance hypothesis. Proc. Nat. Acad. Sci. (USA) 96:1463–1468.

Yodzis, P. 1988. The indeterminacy of ecological interactions as perceived through perturbation experiments. Ecology 69:508–515.

Yodzis, P. 1993. Environment and trophodiversity. In Species diversity in ecological communities: historical and geographical perspectives, ed. R. E. Ricklefs and D. Schluter, 26–38. Chicago: University of Chicago Press.

Young, T. P., J. M. Chase, and R. T. Huddleston. 2001. Community succession and assembly: comparing, contrasting, and combining paradigms in the context of ecological restoration. Ecol. Restoration 19:5–18.

Yu, D. W., and H. B. Wilson. 2001. The competition-colonization trade-off is dead; long live the competition-colonization trade-off. Am. Nat. 158:49–63.

Yu, D. W., J. W. Terborgh, and M. D. Potts. 1998. Can high tree species richness be explained by Hubbell's null model? Ecol. Letters 1:193–199.

Zhang, D. Y., and K. Lin. 1997. The effects of competitive asymmetry on the rate of competitive displacement: how robust is Hubbell's community drift model? J. Theor. Biol. 188:361–367.

INDEX